Satellite and Terrestrial Hybrid Networks

Series Editor
Michel Diaz

Satellite and Terrestrial Hybrid Networks

Pascal Berthou
Cédric Baudoin
Thierry Gayraud
Matthieu Gineste

iSTE

WILEY

First published 2015 in Great Britain and the United States by ISTE Ltd and John Wiley & Sons, Inc.

ISTE Ltd
27-37 St George's Road
London SW19 4EU
UK

www.iste.co.uk

John Wiley & Sons, Inc.
111 River Street
Hoboken, NJ 07030
USA

www.wiley.com

Library of Congress Control Number: 2015944962

British Library Cataloguing-in-Publication Data
A CIP record for this book is available from the British Library
ISBN 978-1-84821-541-2

Contents

Acknowledgments

This book is the outcome of a collaboration between researchers and engineers working in the field of satellite telecommunications. Many of the works cited in the book come from research projects funded by the European Union (SATIP6 and SatSix projects), by the French government through the AESE competitiveness cluster (Ourses project), as well as several studies funded by the National Center for Space Studies (CNES), all of whom I would like to thank for their support. Four theses have been supported during the writing of this book and have made significant contributions to it.

Contributions to this book

Chapter 1:

Patrick Gélard (CNES)
Emmanuel Dubois (CNES)

Chapters 2 and 3:

Olivier Alphand (INP-Ensimag)

Chapters 4 and 5:

Cédric Baudoin (Thales)
Frédéric Nivor (Thesis)
Baptiste Jacquemin (Thesis)

Chapter 6:

Fabrice Planchou (Airbus Defense and Space)
Ihsane Tou (Algerian Space Agency)
Lionel Bertaux (Thesis)

Many thanks to them,

Pascal BERTHOU

Foreword

Once the concept of integrated service appeared with integrated services digital networks (ISDN), the growing adoption of Internet-based technologies into our daily lives has given rise to the concept of "convergence". This is the idea of the progressive merging of information technology, telecommunications and audiovisual technologies into a new and continuously more integrated industry, making the access to the information system more intuitive and easy to use.

Indeed, it appears that the emergence of the digitization and bundling of information was the driving force behind various types of convergence. These range from different uses to diversified points of access, channeling services and networks towards a virtualization of physical infrastructures. Today, digital information flows exchanged via telecommunication infrastructures have become such that it is no longer possible to distinguish an exchange of digital information from a telephone conversation or a message containing audiovisual multimedia content.

This shockwave, which continues to increase the flexibility of the various layers of the information and communication systems, inevitably leads to changes in the structure of economic models, causing significant alterations in the value chain between telecommunications, audiovisual and information industries.

The world of satellite communications has not been spared the effects of this upheaval, and to respond to these changes, stakeholders in the "satcom" sector have been examining the interconnection of satellite radio communications with new information and communication technologies.

This is the subject which this book will address, providing a highly didactic study of the various technical challenges involved in achieving the most transparent integration possible on the different fronts of fixed, mobile and broadcast services with satellite positioning as a core network as well as an access and local network.

The infrastructure for satellite telecommunication has needed to evolve in order to carry various types of traffic and be compatible with frequently updated service offers in an increasingly competitive context. Obviously, the aspect of telecommunication regarding the "quality of service" – looking to offer users the best quality of experience – occupies a special place in this book, and is accompanied by the various technical obstacles to overcome. These include the impacts on the processes for transporting information and communication to mobiles or the new generation of satellite access and diffusion architecture.

This book clearly highlights the various essential aspects to connect satellite communications with the system of new generation terrestrial networks. It provides a detailed insight of a wide range of problems raised by this quest for integration and in doing so represents a unique source of information for those who need to understand the technical challenges which satellite communication networks need to overcome.

The authors, all members of the Internet generation, have spent years actively contributing, through their research, to producing solutions to these technical problems which have influenced the development of the technologies they describe.

Patrick GÉLARD
July 2015

List of Acronyms

2G	second generation (GPRS and EDGE)
3G	third generation (UMTS, HSDPA, HSDPA+ and LTE)
3GPP	third generation partnership project
4G	fourth generation (LTE-advanced)
AAA	authentication, authorization and accounting
ACK	acknowledgement
ACM	adaptive coding modulation
ACQ	(DVB-RCS) acquisition
ADSL	asymmetric digital subscriber line
AF	assured forwarding
AF	(IMS) application function
ANI	application-to-network interface
AP	access point
API	application programming interface
AR	access router
ARC	active resource controller
ASPs	application service providers
AVBDC	(DVB-RCS) absolute volume based dynamic assignment
BA	binding acknowledgement
BACK	binding acknowledgment
BB	bandwidth broker
BBM	break-before-make
BDP	bandwidth delay product

BE	best effort
BER	bit error rate
BSM	broadband satellite multimedia
BU	binding update
C-BGF	core border gateway function
CCSDS	Consultative Committee on Space Data Systems
C2P	connection control protocol
CMT	(DVB-RCS) correction message table
CN	correspondent node
CNES	Centre National d'Études Spatiales
CoA	care-of–address
COPS	common open policy service
COPS-DRA	common open policy service – DiffServ resource allocation
COPS-PR	common open policy service – policy provisioning
CoT(i)	care-of-test (init)
CPE	customer premises equipment
CPM	continuous phase modulation
CR	capacity request
CRA	(DVB-RCS) continuous rate assignment
CSC	(DVB-RCS) common signaling channel
CSCF	(IMS) call/session control functions (P-proxy; S-server; I-interrogating)
CSS	cascading style sheets
CTCP	compound TCP
CWND	congestion window
DAD	duplicate address detection
DAMA	demand assignment multiple access
DCCP	datagram congestion control protocol
DIAMETER	double RADIUS
DNS	domain name server
DSCP	differentiated services code point
DSM-CC	(MPEG2) digital storage media – command and control
DULM	data unit labeling method
DVB-RCS	digital video broadcasting – return channel via satellite
DVB-S	digital video broadcasting – satellite
DVB-S/RCS	digital video broadcasting via satellite/return channel via satellite

ECN	explicit congestion notification
EF	expedited forwarding
E-LSP	EXP-inferred-PSC LSP
eNodeB	evolved node B (LTE)
EPC	evolved packet core (LTE)
ES	(MPEG2) elementary stream
ESA	European Space Agency
ETSI-TISPAN	ETSI-Telecommunications and Internet converged Services and Protocols for Advanced Networking
FBACK	(FMIP) fast binding acknowledgment
FBU	(FMIP) fast binding update
FCA	(DVB-RCS) free capacity assignment
FCT	(DVB-RCS) frame composition table
FEC	forwarding equivalence class
FMIP	fast handover mobile IP
FSS	fixed satellite service
FTP	file transfert protocol
GEO	geostationary orbit
GGSN	gateway GPRS support node
GIST	general internet signaling transport
GPRS	general packet radio service (2.5G)
GSE	generic stream encapsulation
GSM	global system for mobile communications
GTP	GPRS tunneling protocol
GW	Gateway
HA	home agent
HACK	(FMIP) handover acknowledge
HDLB	hierarchical dual token bucket
HHO	horizontal hand-over
HHHO	hybrid HHO
HI	hand-over initiate
HLS	(DVB-RCS2) higher layer satellite
HMIP	hierarchical mobile IP
HNP	home network prefix
HoA	home address
HoT(i)	home test (init)

HSS	(IMS) home subscriber server
HTB	hierarchical token bucket
HTTP	hypertext transfer protocol
I-PEPs	interoperable – performance enhancing proxies
IANA	internet assigned numbers authority
ICMP	internet control message protocol
ID	identifier
IEs	information elements
IETF	Internet Engineering Task Force
IMS	IP multimedia subsystem
INAP	interactive network access operator
INT	(MPEG2) IP/MAC notification table
IP	internet protocol
IP-TV	IP television
ISDN	integrated services digital network
ISP	internet service provider
IST	information society technologies
IT	information technology
ITSPs	internet telephone service providers
ITU	international telecommunication union
ITU-T	ITU telecommunication standardization sector
L-LSP	label-only-inferred-PSC LSPs
LAN	local area network
LBU	local binding update
LCoA	(HMIP) on-link care-of–address
LFN	long fat network
LLS	(DVB-RCS2) lower layer satellite
LMA	(PMIP) local mobility anchor
LSP	label switching path
LSR	label switching router
LTE	long-term evolution
MAC	medium access control
MAG	(PMIP) mobile access gateway
MAP	mobility anchor point
MBB	make-before-break
MF-TDMA	multiple frequency-time division multiplexing access

MIB	management information base
MIP	mobile IP
M2M	machine-to-machine
MME	mobility management entity
MMT	(MPEG2) multicast mapping table
MMUSIC	multiparty multimedia session control
MN	mobile node
MPE	multiple protocol encapsulation
MPEG	moving picture experts group
MPEG2-TS	moving picture experts group – transport stream
MPLS	multiprotocol label switching
MSPs	multicast service providers
MSS	mobile satellite service
NACFs	network attachment control functions
NAR	new access router
NAT	network address translation
NCC	network control center
NCoA	new care-of-address
NCR	(DVB-RCS) network clock reference
NE	network element
NFC	near field communication
NGA	next-generation access
NGN	next-generation network
NIT	(MPEG2) network information table
NMC	network management/operation center
NSIS	next step in signaling
NSLP	NSIS signaling layer protocol
NTLP	NSIS transport layer protocol
OBP	on board processing
OS	operating system
OSI	open systems interconnection
OWD	one way delay
PAR	previous access router
PAT	(MPEG2) program association table
PBA	(PMIP) proxy binding acknowledgment
PBNs	policy based networks
PBU	(PMIP) proxy binding update

P-CSCF	see CSCF
PC	personal computer
PCoA	previous care-of-address
PCIM	policy core information model
PCRF	(LTE) policy and charging rules function
PDCP	packet data convergence protocol
PDF	policy decision function
PDN	packet data network
PDP	policy decision point
PEP	performance enhancing proxy
PEP	policy enforcement point
PES	(MPEG2) packet elementary stream
PHB	per hop behavior
PHoA	previous home address
PIB	policy information base
PID	(MPEG2) packet identifier
PMIP	proxy mobile IP
PMT	(MPEG2) program map table
PrRtAdv	proxy router advertisement
PSI	(MPEG2) program and service information
PUSI	(MPEG2) payload unit start indicator
QNF	QOS NSIS forwarder
QNI	QOS NSIS initiator
QNR	QOS NSIS responder
QoS	quality of service
RA	random access
RA	(IPV6) router advertisement
RACFs	resource and admission control functions
RACS	resource and admission control system
RADIUS	remote authentication dial-in user service
RAN	radio access network
RBDC	(DVB-RCS) rate based dynamic assignment
RC	request class
RCoA	(HMIP) regional care-of-address
RCS	return channel via satellite
RCST	return channel satellite terminal

RFC	request for comments
RLE	return link encapsulation
RMF	resource management function
RNC	radio network controller
RO	route optimization
RRT	return routability test
RSVP	resource reservation protocol
RTO	retransmission timeout
RTP	real time protocol
RTSP	real time streaming protocol
RtSolPr	router solicitation for proxy advertisement
RTT	round time trip
RT-ViC	real time video conferencing
SAC	satellite access control
SACK	selective acknowledgment
SAP	session announcement protocol
SCF	service control function
SCPSs	space communications protocol specifications
SCPS-TP	SCPS transport protocol
SCT	(DVB-RCS) superframe composition table
SCTP	stream control transmission protocol
SD	satellite dependent
SDP	session description protocol
SDU	service data unit
SEs	signaling entities
SESs	satellite earth stations
SGSN	serving GPRS support node
SGW	serving gateway
SIP	session initiation protocol
SLA	service level agreement
SLF	(IMS) subscriber location function
SLS	service level specification
SMTP	simple mail transfer protocol
SNACK	selective negative acknowledgment
SNDU	sub network data unit
SNMP	simple network management protocol

SNO	satellite network operator
SNR	signal to noise ration
SO	satellite operator
SOAP	simple object access protocol
SP	service provider
SPT	(DVB-RCS) satellite position table
ST	satellite terminal
SVNO	satellite virtual network operator
SYN	synchronize
SYNC	(DVB-RCS) synchronization
TBTP	(DVB-RCS) terminal time burst time plan
TCP	transmission control protocol
TCT	(DVB-RCS) timeslot composition table
TDM	time division multiplexing
TIM	(DVB-RCS) terminal information message
TM/TC	telemetry/remote control
TSAPs	transport service access points
TS	(MPEG2) transport stream
UAC	user agent client
UAS	user agent server
UDLR	unidirectional link routing
UDP	user datagram protocol
UE	user equipment
ULE	ultra lightweight encapsulation
UMTS	universal mobile telecommunications system
UNA	unsolicited neighbor advertisement
UNI	user-to-network interface
URI	uniform resource identifier
UTO	user timeout option
UTRAN	UMTS terrestrial radio access network
VBDC	(DVB-RCS) volume based dynamic assignment
VCI	(ATM) virtual channel identifier
VCM	variable coding modulation
VHO	vertical handover
VoIP	voice over IP
VPI	(ATM) virtual path identifier

VPN	virtual private network
VPN SPs	VPN service providers
VSNs	virtual satellite networks
VSNO	virtual satellite network operator
WIMAX	worldwide interoperability for microwave access
WAN	wide area network
WLAN	wireless local area network

Introduction

The history of communication satellites began over 40 years ago with the launching of Anik 1 in 1972, which is considered as being the first geostationary commercial communication satellite. Since then, systems have evolved constantly in order to offer more than telephone services or television broadcasting. With the advent of the Internet, the concept of broadband satellite communication rapidly emerged, with the aim of providing a high-speed connection at any point on the planet. Subsequently, in the 1980s, the first mobile services appeared (Mobiles Satellite Services) with Inmarsat. These systems initially offered maritime telephonic communications, and then mobile data services.

Satellite systems have unquestionable qualities: mainly an extensive geographical coverage for a lower infrastructure cost, with fixed or mobile stations, and a capacity for carrying out large-scale broadcasting. Numerous steps forward in coding and antennae now offer higher speeds. However, satellite communication systems are no longer considered as competitive when compared to terrestrial communication systems. The economic model targeted today is a hybrid of terrestrial networks with satellite connections to supplement them in areas where they are inefficient or lack cost-effectiveness: remote areas and large-scale mobility. Furthermore, the satellite is a suitable medium for the coverage of white zones and offers one of the rare methods of communication which can handle large-scale mobility at high speeds (typically required for plane and train services, etc.)

Convergence is one of the key issues for next-generation telecommunication networks (NGNs). It is also one of the foundations of 4G or 3G long-term evolution (3G-LTE), since it consists of both the convergence of services and fixed-mobile convergence.

This strong trend has given rise to a paradigm shift in order to implement quality of service (QoS) policies in a context where multimedia applications with various demands can be used via different access networks. These QoS policies must, therefore, bring together significantly different QoS management structures depending on the network in question (access or core) while enabling an optimization adapted to each of these networks and services with varied demands. It should be remembered that the current architectures implement a very partial view of QoS from start to finish, and that the solutions implemented at different levels are far from optimal.

This book aims to provide the keys for a successful integration of satellite systems with next-generation terrestrial networks. Digital video broadcasting – return channel via satellite (DVB-S/RCS) family systems (DVB-S/RCS and its evolutions), which are satellite communication systems currently offering the most up-to-date architecture and services – will be used to illustrate the challenges to overcome in order to ensure a successful integration. Of course, the concepts addressed are general and can be applied to other systems, including other rival satellite communication systems.

The presentation of this issue is built around an approach which removes the complexity involved in terrestrial and satellite communication systems. Therefore, this book offers a high-level vision focusing on the components of these systems and their interactions. It is thus aimed at a wide readership, from the designer of the satellite system to the network operator looking to incorporate a satellite option into their portfolio, and from institutional regulators to students wishing to address the issue of terrestrial/satellite hybrid systems.

The various ways of integrating the satellite systems into terrestrial networks will be addressed using several scenarios with different levels of complexity. The management issues related to QoS in terrestrial and satellite networks as well as solutions enabling interoperability will also be addressed. Mobility architectures and their performance will then be tackled. The higher levels will also be addressed with a focus on the role of the transport layer in a hybrid network. All the solutions provided in this book have been developed and tested in a number of European and French research projects. The results were obtained either by measures taken from existing systems, or by realistic imitation platforms, or by the use of simulators when no other option was possible.

Plan

Chapter 1: Satellite and Terrestrial Hybrid Networks

The success of satellite communication systems mainly lies in their wide coverage and reduced time-to-market. Although niche markets, such as ocean and airspace coverage, will continue to exist, the future of satellite systems looks very different. The integration of satellites into terrestrial systems is now the only way to provide a complete offer of fixed and mobile services, with or without broadcasting. This chapter offers a number of hybrid scenarios. These scenarios, known as "tightly coupled", "gateway" or "loosely coupled", will be examined and their impact on the architecture and services will be described.

Chapter 2: Quality of Service on Next-Generation Terrestrial Networks

The QoS guarantee is the cornerstone of the next-generation networks including satellites, in order to remain competitive and profitable. This chapter looks at the essential communication architecture which provides an advanced management of the QoS. Internet engineering task force (IETF) and ITU-NGN approaches will be compared.

Chapter 3: Quality of Service in DVBS/RCS Satellite Networks

DVB-S/RCS is one of the most powerful and flexible satellite communication systems in managing the QoS. This chapter presents the standard DVB-S, its return channel via satellite (RCS) and the recent evolutions of this standard. Particular attention will be given to the QoS architecture promoted by the European Space Agency and the SatLabs group.

Chapter 4: Integration of Satellites into IMS QoS Architecture

The implementation of an integrated QoS architecture, compatible with terrestrial and satellite networks, is a significant challenge. After the presentation of various approaches in Chapter 3, this chapter examines an example of a successful integration in the IP multimedia subsystem (IMS) architecture.

Chapter 5: Inter-system Mobility

Mobility is one of the triggers of business in modern communication networks and must be taken into account in a satellite/terrestrial hybrid system. This chapter gives an introduction to the classification of mobility and Internet protocols. It will then highlight the difficulties and performance problems linked to these hybrid networks. Based on our experience, we will offer recommendations for the management of mobility in these systems.

Chapter 6: The Transport Layer in Hybrid Networks

The transport layer has always provoked debate in satellite systems, although there is now a consensus around proxies (policy enforcement point (PEP)) solutions for improving performance. Hybrid networks have given rise to new problems, such as the severe variation in delay and speed when a mobile changes from one type of network to another, which has a significant impact on the performance of the transport layer. This chapter summarizes the work carried out over the last few years on the transport layer in satellite systems and addresses the issues raised by this layer in hybrid systems. The new perspectives offer by recent evolutions in the Transmission Control Protocol (TCP) protocol will then be evaluated and discussed.

1

Satellite and Terrestrial Hybrid Networks

1.1. Designing satellite and terrestrial hybrid networks

The satellite is a suitable medium for filling in white zones and gray areas due to its wide coverage and accessibility from areas which are not covered by terrestrial infrastructure.

The development of very high-speed access has led to the emergence of new services and uses, which are more and more frequently based on highly demanding audiovisual media for communication purposes. In the short and medium term, in a context where information and communication technologies are assuming increasingly important positions in all sectors and in people's daily lives, it is becoming vital for telecommunication operators to be aware of improvements to existing services on the networks and to have the flexibility to rapidly integrate such new services made possible by this very high-speed access.

The consequence of these efforts is that, for the satellite telecommunications operator, it is necessary to create hybrid systems to forge a convergence between broadcast networks and bidirectional satellites (for fixed and mobile services) on one hand, and terrestrial networks on the other hand, in order to provide higher quality and more transparent access with greater coverage for applications and services which are increasingly demanding in terms of network resources.

The first challenge to overcome with regard to these issues involves the "system" and the need to integrate an effective architecture which takes account of the convergence between satellite and terrestrial networks in an optimized and transparent way (to ensure the delivery of services).

Next-generation networks (NGNs) and next-generation access (NGA) specifications have provided such a convergence by using packet-switching (Internet Protocol (IP), multiprotocol label switching (MPLS), Ethernet, generic stream encapsulation (GSE)/return link encapsulation (RLE), etc.) as a means of interconnection. They look to eliminate the barriers between the various heterogeneous networks by connecting the services between themselves in a secure and accessible way. This is done by using different types of fixed or mobile access terminals, regardless of the underlying transport network. Services can, therefore, be generalized over all types of networks.

This convergence has an impact on the entire value chain and therefore on all of the various stakeholders involved, including service providers, network providers, access network providers, satellite operators, home networks and the terminals of the end user.

Modifications are necessary at different levels of the open systems interconnection (OSI) model and technological challenges need to be overcome. Several hybrid scenarios must be considered in which the satellite can play a technically, economically and socially useful role.

1.2. Hybrid scenarios

Over the last decade, a number of new access network technologies emerged for access to Internet services. At the same time, cellular networks, initially designed for mobile telephone and voice services, evolved to offer more advanced services and above all Internet access.

Moreover, the progress of mobile terminals (mobile telephones, smartphone, ultrabooks or laptops), whose size and weight have been significantly reduced, incorporate increasingly wireless network interfaces (3G/4G, WiFi, Bluetooth, near field communication (NFC), etc.) and communication capacity. These wireless communication technologies (WIMAX and 3G/4G-LTE) have given the user the ability to connect to services from anywhere, anytime, while enabling mobile Internet access.

This trend is so strong that the offer of an “always on” service is now one of the requirements for the design of new network infrastructures.

The NGN and 4G concepts are entirely in line with this approach. Services or applications are designed with no specific type of access network in mind (wireless, cellular, cable, optic, etc.), but are based on core IP technology, which is now a cornerstone of convergence between telephone and data services.

In NGNs or 4G networks, the "always on" paradigm is intended to give general mobility to service users, with a fully transparent change of access network as long as it is compatible.

Application, as well as the underlying protocols, must be consistent with the changes in networks. However, this requirement remains very ambitious since the networks are heterogeneous and potentially operated by a wide variety of stakeholders. Therefore, a number of economic (business, role model, etc.) and technical (quality of service (QoS), authentication, authorization and accounting (AAA), security, etc.) questions arise.

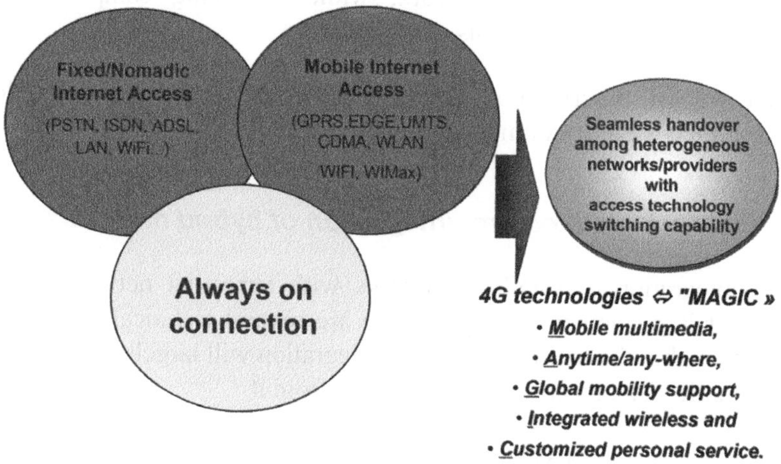

Figure 1.1. *Trends with 4G/NGN*

Therefore, it is crucial for satellite systems to follow this trend and demonstrate their compatibility with NGN/4G networks. This is of utmost importance for the satellite broadband market. Indeed, most stakeholders (industries, suppliers and research laboratories) are arguing for the integration of satellites into this architecture.

There are a number of cases where satellite/terrestrial hybrid networks would be particularly advantageous. As a supplement traditional terrestrial access technology, satellite systems offer a real benefit for mobile users, and in a more general sense, for the deployment of mobile networks. Satellite networks offer extremely wide coverage and a high accessibility rate, with capacity in terms of performance, QoS management or security, which is entirely comparable to traditional networks. Of course, satellite networks will not compete with terrestrial networks, but can supplement their coverage and offer an alternative solution which can be very useful when terrestrial infrastructure becomes ineffective (mass congestion and attack) or is destroyed (natural disaster), or is simply not available (no coverage).

Therefore, the typical use would now be mainly in the sectors of civil protection, military (theater of operations) or transport (maritime, aeronautical, railway, etc.).

The following section analyzes the impact of these uses on the overall architecture of the hybrid network, while remaining compatible with NGN/4G architectures (protocols and standards).

First, we will describe how the integration at the system level could be done, and then we will examine various scenarios.

1.2.1. *Network architecture: integration of hybrid networks*

The integration of satellite networks with terrestrial networks can be carried out in a number of ways. There are several technical solutions for this problem, but the main criterion for integration will largely be dictated by the role models and businesses which come from it.

Nevertheless, it is possible to define three generic types of integrations:

1) *Tight coupling integration*, where the mobile system (3G, long-term evolution (LTE) and WIMAX) is extended to use the satellite as an alternative access channel in a completely transparent manner.

2) *Gateway integration*, where the satellite is integrated into the infrastructure of the mobile network, not directly at the level of the air interface, but via a specific gateway enabling access to the core mobile infrastructure.

3) *Loose coupling integration*, where a specific satellite system interface is added to the mobile satellite terminal in order to enable access to a terrestrial IP network via this interface. Multimodal and multitechnology

terminals capable of generating several interfaces and their specific protocols (e.g. DVB-RCS+M) are, therefore, necessary.

These three scenarios are described in the following sections to provide more technical details.

1.2.2. *Tight coupling integration: an integrated approach*

With the "tight coupling integration" approach, the satellite is completely merged into the targeted mobile system (3G, LTE and WIMAX), in a transparent manner for the mobile user. The radio access interface is extended (infrastructure and protocols) in order to integrate a satellite channel as an alternative access interface for the mobile user.

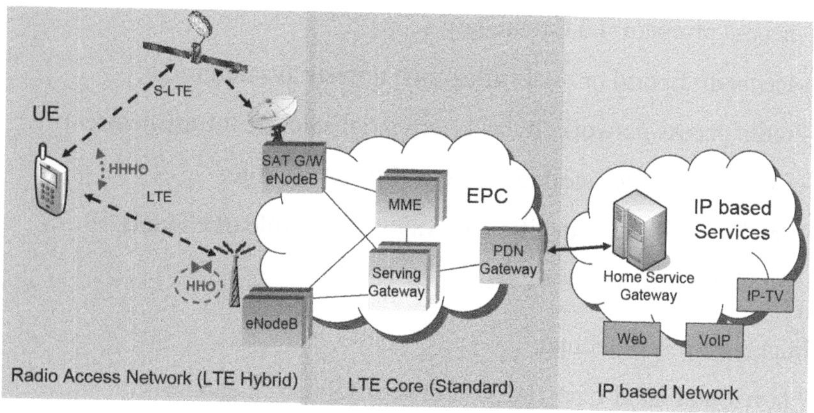

Figure 1.2. *Tight coupling architecture*

The *Centre National d'Études Spatiales* (CNES), or national centre for space studies, is currently carrying out studies based on this approach, most notably as part of the SWIMAX project. If we examine the LTE system, the satellite would be directly integrated into the core infrastructure and the gateway satellite would become a standard interface (an enodeB). The mobile terminal can communicate with the gateway satellite via a channel satellite by using traditional terrestrial protocols (which can, however, be adapted to work on a channel satellite).

This approach is considered as the final step in the integration of the satellite into the hybrid network. The satellite system is designed to be fully compatible with the mobile protocols and is fully integrated into the core network via a standard interface eNodeB. This is also the most powerful approach from a user's point of view.

Unfortunately, it is also the most complex since it requires very powerful hardware performance in order to maintain the small size of the portable equipment.

The terminal is hybrid since it can interact with the satellite or a terrestrial antenna by using the same protocol stack (LTE or WIMAX). The management of mobility is handled by these protocols like in a traditional terrestrial cellular network.

The characteristics of this approach in an LTE network are:

– access protocol: LTE (standard scope);

– terminal: hybrid or dual (integrated terrestrial/satellite);

– radio access network: hybrid (terrestrial/satellite infrastructures);

– satellite: mobile satellite services (MSSs) satellite;

– satellite gateway: specific gateway with the role of enodeB;

– mobility: provided by LTE.

Important points include:

– horizontal hand-over (HHO) between terrestrial enodeBs;

– hybrid HHO (HHHO) between satellite/terrestrial enodeBs;

– the mobility at the network level is ineffective. The IP address is maintained by a single packet data network-gateway (PDN-GW).

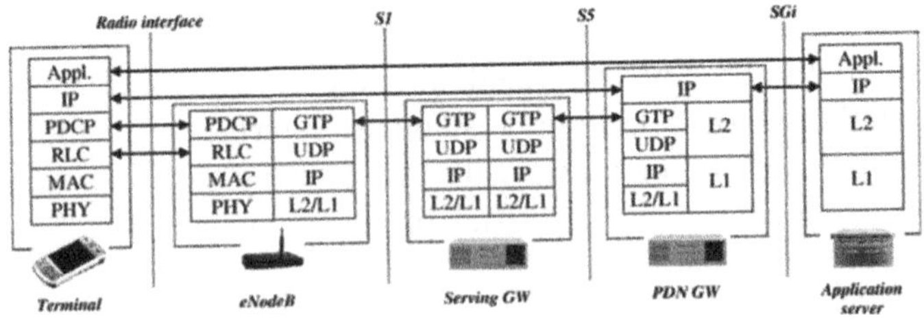

Figure 1.3. *LTE protocol stacks (User Plan – 3GPP standard documents)*

1.2.3. *Gateway integration*

With the "gateway integration" model, the satellite is integrated in the mobile network as a gateway. Indeed, it is not placed at the radio interface with the mobile, but as a specific gateway to allow access to the core of the mobile network.

Therefore, the mobile terminal is a traditional terminal, in compliance with the standard of the mobile network targeted (e.g. LTE or Wimax). This is no longer dual equipment and the satellite interface remains a traditional fixed satellite interface (fixed satellite service (FSS)). The mobile is connected to a traditional eNodeB, which is interconnected with the core network by a satellite link. The satellite network has an interface with the core terrestrial mobile network. In an LTE model, it may fulfill the role of an eNodeB or a serving gateway (SGW).

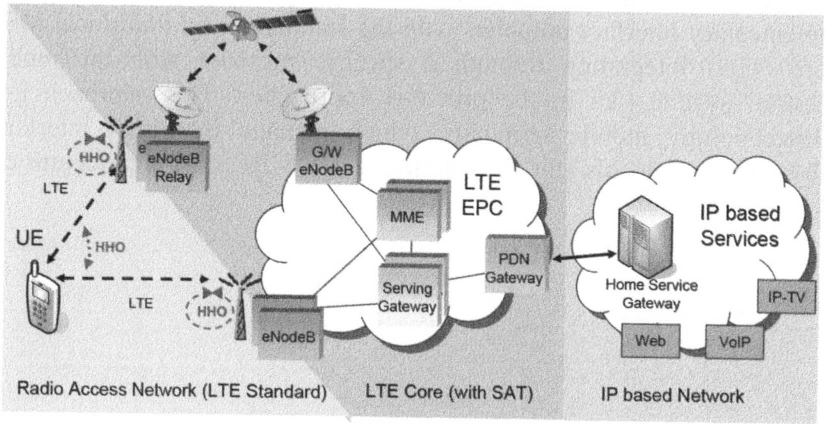

Figure 1.4. *LTE gateway architecture*

The characteristics of this approach in an LTE network are:

– access protocol: LTE (standard);

– terminal: LTE (standard);

– radio access network: LTE (standard with satellite gateway to core);

– satellite: fixed satellite services satellite;

– satellite gateway: standard gateway in the role of an enodeBor SGW;

– mobility: provided by LTE.

Important points include:

– HHO between terrestrial enodeB and gateway eNodeB;

– HHHO between satellite/terrestrial enodeB;

– mobility at network level is ineffective. The IP address is maintained by a single PDN-GW.

1.2.4. *Loose coupling integration*

With "loose coupling integration", a specific satellite interface is added to the terminal to connect to the IP network via a specific access network. It can be differentiated from the first approach by the fact that this time, the supplementary interface complies with the standards of a traditional MSS. There is no integration through a specific protocol with the mobile terrestrial system like in the previous approaches. This approach uses multitechnology mobile terminals, which manage the usual interfaces with specific protocols (for the satellite, a DVB-RCS+M architecture can be used).

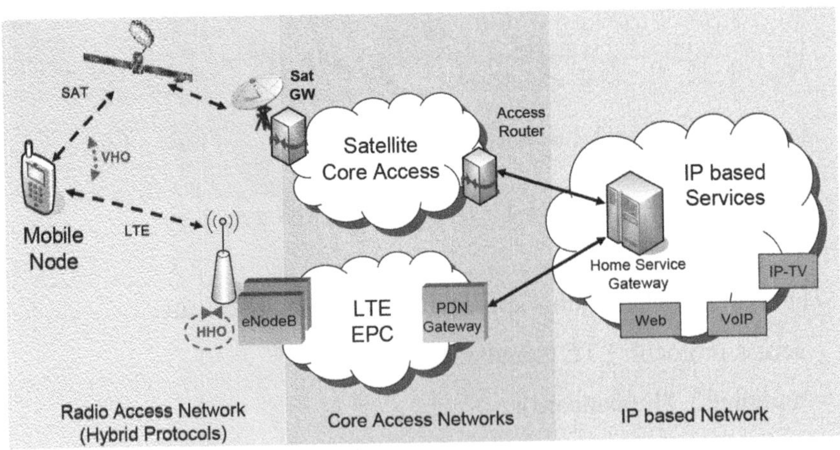

Figure 1.5. *LTE/satellite loose coupling integration*

This architecture can be applied generally to all technologies, and is not limited to LTE. The mobile terminal connects to an IP network via heterogeneous access networks.

1.3. Case study: loose coupling integration

1.3.1. *Use case and user profile*

To define the use case, it is important to define the following points which specify the user's profile:

1) *nature of traffic*: asymmetrical data streams, length of connections, variable flows, encryption if necessary, etc;

2) *nature of geographic mobility:* the distance which a mobile user can be from the home agent;

3) *hand-over frequency:* to be determined depending on the types of underlying access networks.

It is important to be aware of these parameters because they will guide the selection of suitable mobility mechanisms or may be used as requirements in the definition of an appropriate new mechanism.

Nevertheless, this is not an easy task because user profiles can vary greatly according to their needs, zones and types of movement. The nature of the traffic is possibly the least predictable parameter because the uses of the mobile network are very changeable, they have many variable applications, and an operator's service offer is greatly increased in line with its capacity for innovation.

1.3.2. *Proposal of a scenario*

We will pursue the analysis of the "loose coupling" approach. Indeed, although the "tight coupling" and "gateway" approaches are not excluded, we believe that it is interesting to develop a case with a real vertical hand-over (between multiple technologies). In the first two cases, mobility IS directly managed by the LTE and is therefore entirely transparent at the higher layers.

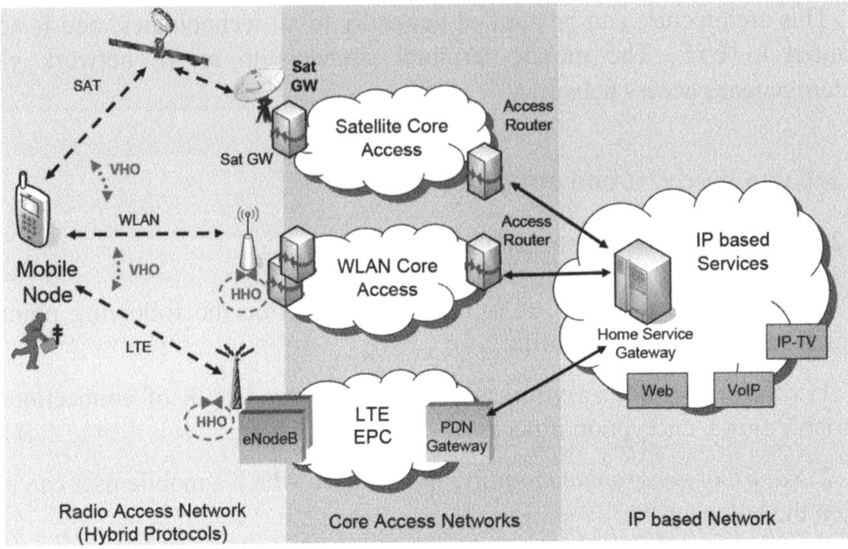

Figure 1.6. *Heterogeneous hybrid architecture for mobile nodes*

The characteristics of this approach are:

– access protocols: multiple, heterogeneous and hybrid protocols;

– terminal: multimodal, adapted to different networks;

– radio access network: terrestrial (LTE or other) and satellite;

– satellite: MSS satellite;

– satellite gateway: standard gateway;

– mobility: horizontal and vertical hand-over.

Important point:

– mobility is managed by the network level and therefore by IP mobility stacks.

Based on the previous architectures, it is possible to address the case of mobile networks. This case is particularly interesting in the sector of public transport, communication vehicles as well as the military. The satellite thus becomes an alternative to the terrestrial networks for access for vehicles. In this case, part of the network is mobile, referred to as "mobile networks".

Therefore, it is the router which manages its mobility, while the nodes it hosts appear as fixed in the rest of the network.

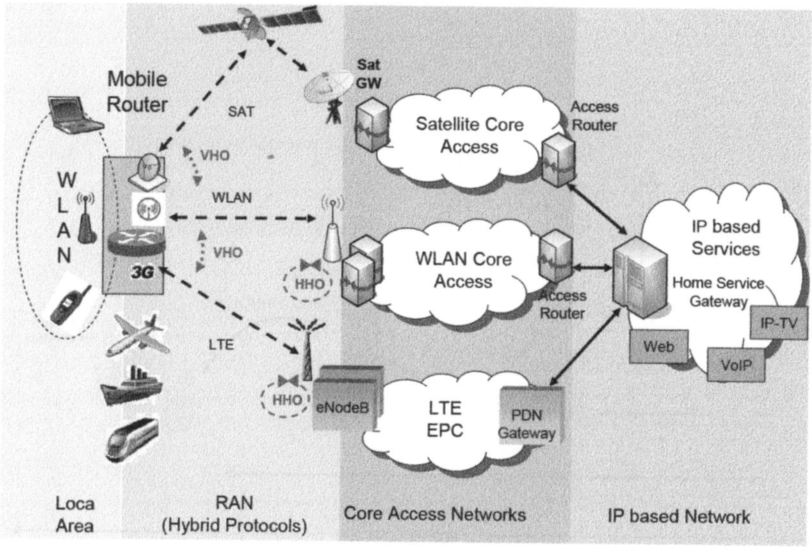

Figure 1.7. *Heterogeneous hybrid architecture for mobile networks*

1.3.3. *Profile of mobile users*

An example of a user profile in this case may be as follows:

– coverage: wireless local network (WLAN) and/or cellular network (3G/4G), supplemented by a wide coverage satellite network. Indeed, handovers are not frequent (vertical handover (VHO) and HHHO), since they will often be managed by access technologies (HHO in 3G or LTE);

– geographic mobility: from short distance (for mobile users) to intercontinental mobility (for transport, e.g. in an airliner);

– nature of the traffic: regular traffic for Internet access, including web navigation (HTTP), email (SMTP), file transfer (FTP) and audio/video streaming (RTP). Except peer-to-peer (P2P), all these protocols generate asymmetrical flows, mainly from the network to the mobile. The length of connections varies (short for HTTP and long for SMTP or RTP). For security reasons, an IPsec-type protocol may be adopted due to its extensive use;

– mobile equipment: mobile terminal or mobile router.

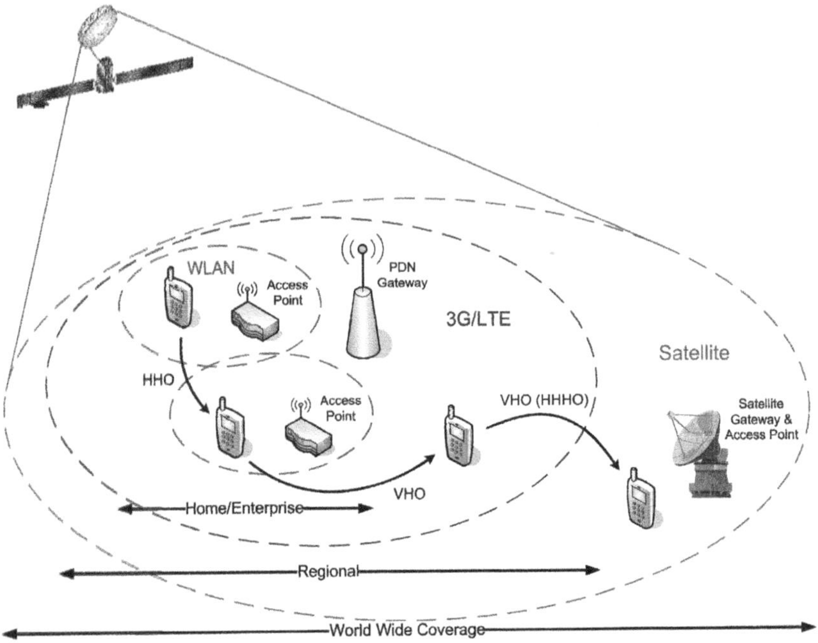

Figure 1.8. *Network coverage in the mobility scenario*

To complete the network integration, the terminals and routers which manage several interfaces must have an algorithm for selecting the "best" network according to the service or application activated by the user. This interface selection may be automatic or based on preferences (profile, cost, etc.) and/or on network availability (transmission capacity, signal level, bit error rate (BER), availability of resources, security, etc.).

1.4. Conclusion

Following this brief presentation of the important aspects of hybrid networks, it would appear that the market and technologies have accepted the integration approach and the process of creating hybrid networks. Indeed, dictated by users' requirements, emerging hybrid networks already exist due to the complementarity offered by WiFi networks and 3G/4G mobile networks. Nevertheless, these approaches may be considered as being completely uncoupled because, although some operators offer services for two types of network (WiFi or 3G/4G), it is rare for these networks to be interoperable, even by a common authentication, and more generally by a general service offer from both technologies.

It is difficult to predict the future of these mobile networks, although some trends are starting to emerge. The widespread distribution of femtocells, integrated into asymmetric digital subscriber line (ADSL) boxes, is encouraging general 3G/4G access at the expense of traditional wireless networks. However, the evolution of the Internet mobility architectures and very high-speed wireless access protocols could, on the contrary, lead to an increase in access via these networks. On the other hand, taking a more pragmatic approach, given the specifics of each access technology in terms of coverage, speed and services, it is highly likely that these technologies will continue to work side-by-side for some time to come. Nevertheless, as has been argued in this chapter, integration solutions for these networks need to be proposed despite the significant challenges which this presents.

The following chapters will focus on the main challenges currently faced in integrating satellites into these future hybrid networks. The first involves the management of the QoS, via QoS management architectures of terrestrial and satellite networks, and also via architectures and standards for new generation hybrid networks.

2

Quality of Service on Next-Generation Terrestrial Networks

Over the last few years, some important work on research and standards has been carried out regarding the implementation of quality of service (QoS) in internet protocol (IP) networks. In this chapter, the standards and studies regarding QoS are presented according to the main two approaches corresponding to the two different visions of the implementation of Information Technology (IT) networks:

1) The internet engineering task force (IETF) approach: IETF offers a set of standards facilitating the implementation of QoS or integrating the requirements of this implementation. The objective of the IETF standards is to provide simple and scalable mechanisms enabling the progressive introduction of QoS into the networks without necessarily setting out a general framework or overall management structure from end-to-end to ensure this QoS.

2) The ITU-NGN approach: ITU involves the implementation of QoS from end-to-end in networks using a single transport system, thus allowing a convergence of voice and data networks. Moreover, the management of QoS in the next-generation networks (NGNs) proposed by ITU must dynamically take account of the requirements of different applications which may be used on this unified network.

2.1. IETF approach

2.1.1. *Network level*

The network level of the Internet (i.e. Internet Protocol) offering a best effort service, without reliable delivery or connection, helped to ensure robustness and wide-scale deployment due to a simple protocol layer involving a similar process for all of the flows crossing the communication network.

The best effort service was initially deployed in IP networks and is currently still predominantly used on the Internet. This service is characterized by a complete absence of guarantees regarding the delivery of IP datagrams to the receiver. Therefore, the service currently offered by this layer does not offer guarantees in terms of QoS. In this initial model, most of the QoS mechanisms are ensured from end-to-end by an optimization of transport and/or application levels.

The following sections will present the various IETF architectures enabling the introduction of QoS guarantees at the level of the IP layer.

2.1.1.1. *IntServ integrated services and the RSVP signaling protocol*

Integrated services [BRA 94a] provide an architecture based on a reservation by flows, which is made up of two main classes of service, as well as the current best effort class. These main two classes can be broken down into subclasses depending on the flow demands requesting reservation. The definition of these classes is as follows:

– the "Guaranteed Service" class defines an upper time-limit for transmission of a set of packets belonging to a stream from this class. Some of the resources of each entity crossed are reserved for the stream in question. The parameters of *latency*, *bandwidth* and *jitter* can, therefore, be guaranteed by this service;

– the "Controlled Load Service" is used for streams requiring an improved service compared to the best effort service.

A signaling protocol, Resource ReServation Protocol (RSVP) [BRA 97], is used along the path from the emitter to the receiver in order to reserve the resources necessary for the stream in the required class. The specification of the stream in an RSVP message contains:

– the class of service required;

– and generally, two types of digital parameter:

- an RSpec-type specification describing the desired QoS, containing the latency and bandwidth necessary,

- a TSpec traffic specification describing the precise characteristics of the stream traffic, including the average flow rate, burst size, peak cell rate and length of data units.

A resource reservation request is sent by the receiver in response to the RSVP sent by the emitter. If the routers crossed by the reservation request have the necessary resources for the stream in question, a setting will be made in these routers in order to identify the stream and reserve the resources to which it corresponds. This resource reservation process is shown in Figure 2.1.

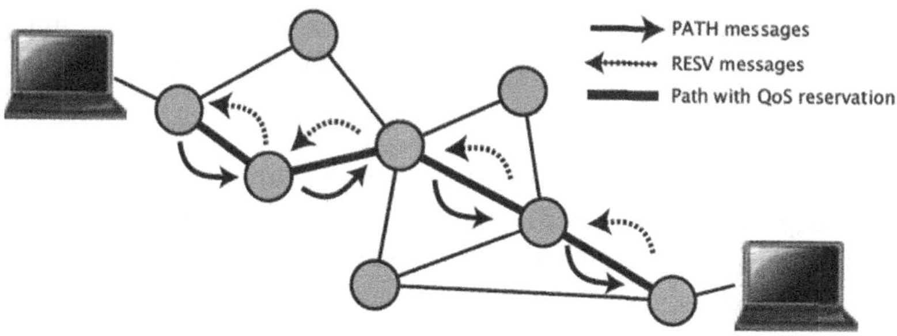

Figure 2.1. *Reservation of resources by RSVP protocol for an Intserv class stream*

An admission control mechanism is necessary to determine the streams which will be accepted depending on:

– the contract between the client and the Internet Service Provider (ISP);

– and the resources available.

There is a *standardized contact* [CAM 03] between the protocols, describing the high-layer sessions (such as Session Description Protocol (SDP)), and these resource reservation protocols. However, this contact is limited to the passage of IP addresses and ports identifying the streams for which resources must be reserved. This information is, therefore, insufficient for accurately characterizing a stream in terms of QoS and ascertaining the resources necessary to reserve them.

The architecture of QoS by stream offered by integrated services helps to provide a strict guarantee of the quality offered to various streams having reserved resources along the path taken. Nevertheless, the question of large-scale deployment of this architecture quickly arose and revealed its limits.

Indeed, the main problem encountered in terms of the large-scale deployment of this architecture rests on the fact that each router which is crossed by a stream must maintain a steady state for this stream. If we consider the number of streams crossing routers, especially in the core network where thousands or even millions of streams are being transported, it becomes apparent that this architecture will produce processing loads which are incompatible with the processing speed necessary in these routers.

Furthermore, complex communication between routers must be introduced to exchange routing information, which adds further traffic load and extra complexity.

The third negative aspect of this architecture is linked to the requirements of its deployment. Indeed, integrated service mechanisms must be implemented over all routers crossed to ensure an end-to-end guarantee. Solutions exist if some routers do not offer the required mechanisms or if one of the routers on the pathway malfunctions, such as the use of IP tunneling, but these solutions add yet more complexity to the architecture as well as a possible element of uncertainty to the QoS provided.

Given the significant limits to this type of approach, IETF created another working group to produce a new architecture to overcome the problems of the previous approach. The architecture resulting from this working group is presented in the following section.

2.1.1.2. *DiffServ differentiated services*

IETF standardized a QoS architecture by class – differentiated services [NIC 98, BLA 98] – in order to offer a simpler large-scale deployment than the one provided by integrated service.

An example of the deployment of this end-to-end architecture of a network is given in Figure 2.2.

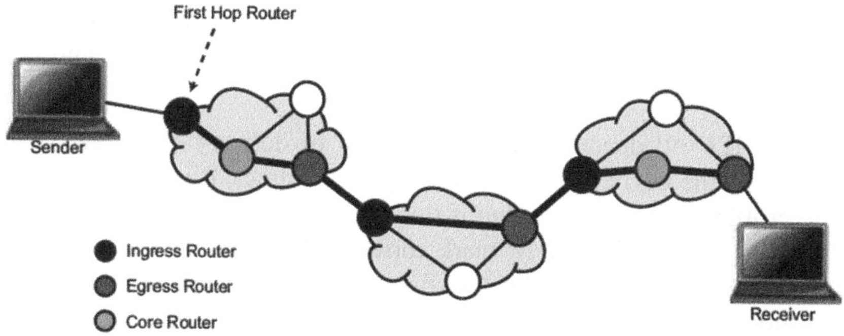

Figure 2.2. *Overview of the DiffServ network*

Unlike integrated services, which offer a stream-oriented QoS architecture, differentiated services take a limited number of service classes to apply to stream aggregates, thus drastically reducing the states to maintain in the routers, especially those in the core network. Indeed, in this case, the number of states is limited to the number of service classes which in a normal setup does not exceed 14 classes (one expedited forwarding (EF) class, 12 assured forwarding (AF) classes and one best effort (BE) class).

The stream classification operations – and therefore the added complexity – are only carried out in ingress routers of the "differentiated service domain" (First Hop Router in Figure 2.2). Stream operations in these routers are more sophisticated than those carried out in core network routers, meaning a potentially significant processing load. However, these routers only need to process a limited number of streams since they are situated at the start of the DiffServ domains. Consequently, they can handle these operations which are more substantial in terms of processing.

These two aspects, which distinguish the architecture of differentiated services from that of integrated services, make large-scale deployment of this type of architecture easier.

This differentiated services architecture involves the use of three main classes of services, including the "Best Effort" class.

Here is a description of these service classes as given by IETF:

– The "EF" class [DAV 02], also known as premium class, is the highest priority class giving streams which use the lowest latency and variations.

This is due to the fact that they are processed as a priority in the routers which implement this service and therefore have a short waiting time in the router queues.

– AF [HEI 99] denotes a high likelihood of delivery of packets which comply with the traffic profile defined by the contract. It is possible for this traffic profile to be exceeded, but if this does happen the data which exceed the contract will receive a new mark corresponding to a lower likelihood of delivery. Indeed, they will be the first to be eliminated in the event of congestion in the network. This service class will have priority in the routers below the EF class but above the best effort class.

– The best effort service class will take all the resources not used by the service classes described above and will have the lowest priority in the router queues. This will lead to delays and higher rates of data loss in the event of congestion in the network. It is advised to reserve some resource for this class of traffic in order to avoid a complete blockage of streams using this class of service.

– The architecture of a Diffserv node is shown in Figure 2.3. It consists of an input which includes a classifier whose role is to determine the correct pathway inside the node. The operation of marking streams into a given class, carried out in the border routers, is based on a number of criteria in an IP packet, namely the source and destination addresses and ports, and the protocol ID (multifield classification). This classification assumes that there are rules for redirecting streams toward the appropriate class. A meter determines whether a packet has the necessary performance required by its class and then decides on its processing. The conditioner is responsible for shaping a group (traffic policing, packet dropping, etc.) so that they comply with the given profile. The packets are then redirected toward the scheduler queues.

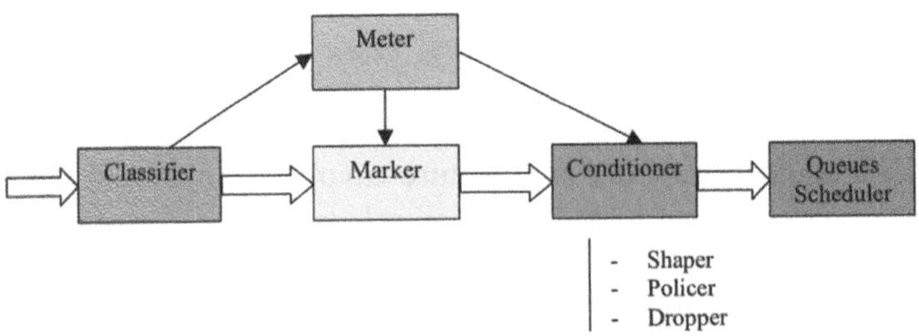

Figure 2.3. *Logical structure of the classifier and traffic conditioners*

To summarize, within a DiffServ domain, the QoS requirements which can be guaranteed are as follows:

– *EF service class* guarantees *low levels of latency, variation and loss rate* (thus, a high level of reliability).

– *AF service class* guarantees *high reliability* for the data which complies with the profile. It can guarantee a latency and variation which complies with the profile. However, the *reliability* of the traffic *outside the profile* is *not guaranteed* and its routing depends on the congestion in the network.

– The *best effort* service class offers no guarantee in terms of the routing of data. This depends on the congestion in the network.

2.1.1.3. *Conclusions on QoS at the IP level from IETF*

It is clear that there are solutions which can introduce a certain level of QoS into the Internet (IP) layer thus enabling a modification of the services provided by this layer to the requirements of the higher layers, thus leading to an improvement in the quality perceived by applications and their users.

Nevertheless, the deployment of these QoS architectures is not done on a large scale despite there being a need for such a deployment.

A number of reasons can be given to explain this situation:

– the first, more specifically encountered by integrated service architecture, is the difficulty of applying this type of approach on a large scale, as explained previously;

– the second, regarding differentiated services architecture, is linked to the difficulties encountered in deploying this approach between several domains. At the borders of the domains, the links and traffic engineering to be implemented can be complex;

– a third reason which can be put forward is the absence or shortage of standards to help to propagate QoS information between the layers and throughout the network. This information would first be necessary to allow the IP layer to configure the QoS in accordance with the requirements expressed by the higher layers. They would also be useful in creating an end-to-end configuration.

This principle of information exchange between the layers of the model in order to take decisions regarding the QoS is not properly addressed in the

IETF architectural proposals. As we have seen, the standard classification of streams is carried out without taking account of the QoS requirements of the higher layers, but only by elements allowing the identification of packets belonging to the same stream. A preliminary step for selecting streams to be marked must, therefore, be carried out, but it is not standardized by the IETF.

The differentiated service approach proposes an intermediate solution, the *bandwidth broker (BB)* [NIC 99], which memorizes the state of the admissions and dynamically accepts or refuses new traffic requests based on policies and the availability of resources in the various classes. The BB implements and initiates an admission control algorithm for deciding whether to admit new streams into the network. In its domain, the BB corresponds to a policy decision point (PDP) by deciding to admit or refuse a stream while border routers, for example, take the role of policy enforcement points (PEPs), by applying marking, scheduling and queue management policies for data stream requirements. To implement an end-to-end allocation of resources via different administrative domains, a BB, which manages a specific domain, needs to communicate with the BBs in adjacent domains.

The BB, therefore, enables the dynamic exchange of information upon the identification of streams whose QoS needs controlling. Nevertheless, this approach is limited to the requests requiring a prioritization of streams. Therefore, it allows higher layers to communicate with the network layer but does not resolve the problem of the expression of requirements and connecting them with the services available in the network layer.

The deployment of the DiffServ architecture between several autonomous domains to ensure a QoS guarantee to satisfy the requirements of next-generation applications requires service level agreements (SLAs) between clients and service providers. These agreements must be passed on to adjacent domains and between a service provider and an end user. This issue is addressed in section 2.1.1.5.

2.1.1.4. *MPLS: multi-protocol label switching for the application of quality of service*

Currently, routing in IP networks is based on the destination of packets. Indeed, the decision to route a packet at a network node relies solely on the destination address. Therefore, it is impossible to carry out this routing so as to balance the load in the network.

In order to offer an alternative to this routing based on destination, a technique using packet labeling was put forward by the IETF in 1997: MPLS [AWD 99]. MPLS is a mechanism enabling the routing of packets according to the label which they are given upon entering an MPLS domain (an administrative domain) and which is taken away from them when they leave this domain. Within the MPLS domain, the routers are not based on the destination IP address, but solely on the MPLS label. Therefore, the packets are divided into forwarding equivalence classes (FECs). All the packets in the same FEC, with the same MPLS label, follow the same network pathway in the MPLS domain. The allocation of a label, and thus an FEC, to a packet is defined by using a number of different fields, such as source and destination IP addresses, the protocol number in the IP header and can also be based on the port numbers in the transport layer. The labels can, therefore, be given according to the application in question, and possibly according to existing service level specifications (SLSs). In an MPLS domain, routers able to read MPLS labels are label switching routers (LSRs). Distinction is made between the LSRs in the core and at the borders of the MPLS domain (see Figure 2.4). The remaining LSRs crossed by packets from a given FEC between the input and output LSRs in the MPLS domain are known as label switched path (LSP).

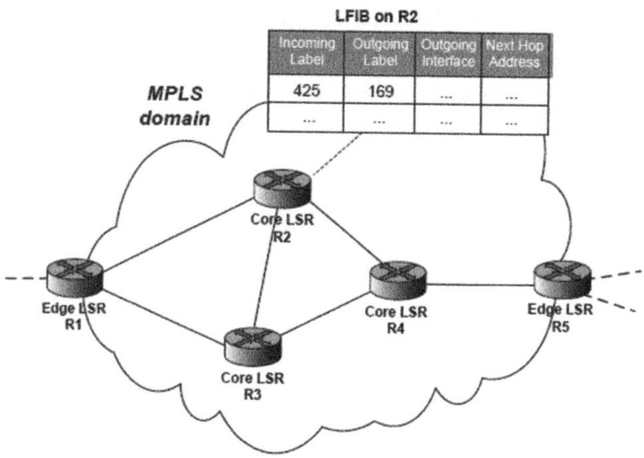

Figure 2.4. *Example of an MPLS domain*

The MPLS label is made up of several fields, as can be seen in Figure 2.5: a "Label" field of 20 bits which encodes the MPLS label, an experimental field, "exp", of 3 bits which can encode a field in accordance with the DiffServe differentiated services code point (DSCP), an "S" field of 1 bit and an 8 bit "Time-to-Live" field.

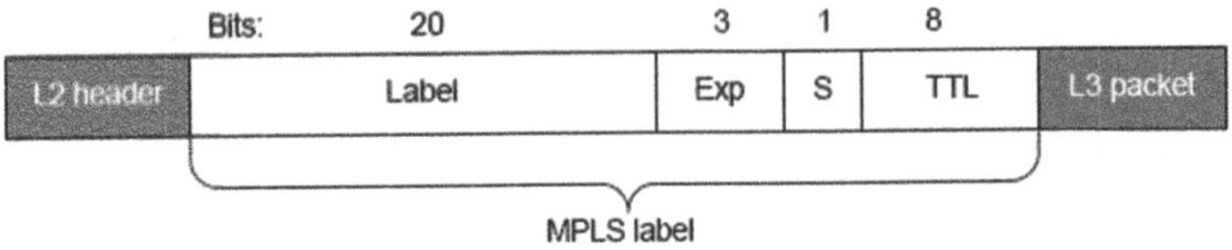

Figure 2.5. *MPLS field*

Implementation of the quality of service with MPLS

MPLS traffic engineering is a means of controlling congestion which traffic can encounter in the network domain. However, MPLS is not aware of the base of service classes required for the traffic, and therefore cannot provide a direct support for QoS. Indeed, in an MPLS domain, once an IP packet, in an input LSR, is encapsulated with the MPLS header, the routing decisions are only based on the label contained in this header, and the IP header is no longer examined.

For this reason, the key problem, so that the MPLS routing complies with the class differentiation proposed by the DiffServ architecture (see section 2.1.1.2), is to connect the DiffServ service classes with the LSP.

Le Faucheur *et al.* [LEF 02] describe the necessary mechanisms for the DiffServ support in an MPLS domain. Two solutions, corresponding to two types of LSP, have been offered: EXP-inferred-PSC LSP (E-LSP) and label-only-inferred-PSC LSPs (L-LSP):

– E-LSP: in this case, the service class to apply for routing is directly encoded in the 3 bits "exp" of the MPLS header. This field thus includes the service class and the likelihood of a packet being dropped (drop precedence: AF11, AF12 and AF13). When a packet arrives at an LSR input router, it is first classified into the corresponding FEC class, which determines the label in the MPLS header. Secondly, the IP service class is determined by examining, for example, the DSCP field of the IP header, assuming that this has already been allocated. When a packet arrives at a core LSR router, the following node is always determined in accordance with the label in the IP header. However, once the packet has been transmitted to the output interface of the router, the queue in which it has been memorized and its drop precedence are determinded according to the "exp" fields of the MPLS (in relation to the IP service class);

– L-LSP: for this second solution, the packet sequencing is linked to the MPLS label itself. The MPLS label is, therefore, linked to the DiffServ service class, and to the sequencing of the packets. However, extra information is necessary to obtain the likelihood of packet dropping in a given service class. In this case, this information is included in the "exp" field of the MPLS header. For instance, the packets belonging to the AF1 service class will all have the same MPLS label, but the packets in this class with different drop precedences (that is to say AF11, AF12 and AF13) will be marked due to the "exp" field in the MPLS header. The MPLS label cannot be determined in advance because it includes the information on the service class. Therefore, a signaling protocol such as RSVP must be used when the LSP pathway is implemented so as to determine the connection between these two parameters.

2.1.1.5. *Management of SLAs and SLSs in the QoS context defined by the IETF*

In order to provide QoS guarantees which can be ensured by the communication network to satisfy the requirements of next-generation applications (especially, multimedia applications), SLAs made between clients and service providers are the subject of a number of research projects. Above all, two European projects (AQUILA and TEQUILA) have made contributions for defining and implementing these agreements [SAL 00, GOD 00].

The objective of these projects is to produce a framework which is able to introduce a level of QoS corresponding to the new requirements in the current best effort network while optimizing the use of the resources in the network. This work proposes agreements between two domains and agreements between a service provider and an end user (see Figure 2.6). This type of agreement will, therefore, be suitable for the differentiated services architectures described previously which require agreements between different DiffServ domains as well as between the access networks and end users (in order to ensure an end-to-end QoS). Work has been carried out to this end for DiffServ architectures, such as the common open policy service – DiffServ resource allocation (COPS-DRA) protocol [SAL 02]. Indeed, the DiffServ architecture is based on the current division of the Internet into autonomous domains, or groups of nodes which respond to a single administrative authority. Within the "autonomous domain", it is possible to distinguish the border routers (ingress or egress routers which can be directly connected to a user network or another domain) and core routers. Any client (operator or individual) who wants to benefit from domain services needs to negotiate an agreement with the operator in the form of an SLA. This contains the technical and non-technical terms and conditions. We will define the notion of SLA in much detail below.

An *SLA* [WES 01] is an agreement which emerges from a negotiation between a client and a service provider which contains quantifiable information on the service to be delivered by the service provider to the client and the possible penalties to impose if this service is not delivered. The technical part of the agreement is known as the *SLS* [GOD 00]. Above all, it includes the QoS parameters which characterize the services offered.

Figure 2.6 represents the various stakeholders in the network and the agreements which may be made between them:

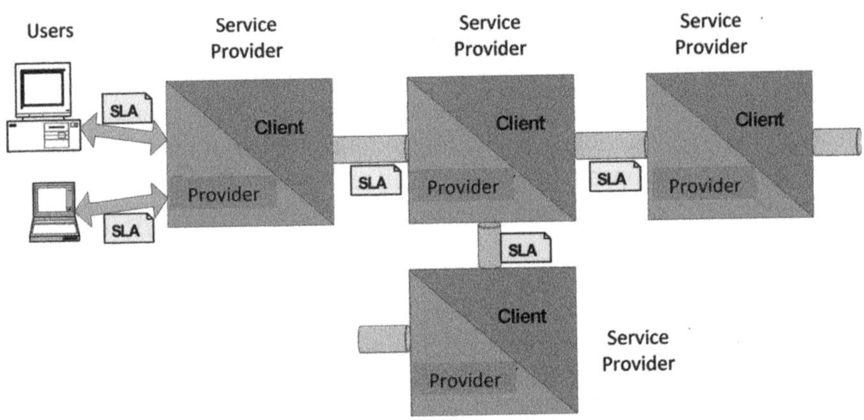

Figure 2.6. *Diagram showing users, service providers and the SLAs negotiated*

This figure shows that network service providers are divided into two distinct groups:

– a provider offering SLAs to end users or to other service providers, which offers network services in its domain depending on reliability;

– a client negotiating services and resources with adjacent domains.

This subdivision of service providers, which are both provider and client, helps to ensure an end-to-end negotiation of agreements and thus services offered to end users.

No standard description of SLAs is currently available. However, a set of generally accepted criteria which must feature in the specification of *SLAs* and in the specification of technical parts of these agreements, the *SLSs*, can be given:

– the respective responsibilities of the client and the service provider;

– details on the procedures to follow in the event of non-compliance with the guarantees made in the SLS;

– the price of the service and the reduction in the event of non-compliance with the agreement;

– reporting to the client on the QoS delivered;

– the possibility of the client to modify the agreement;

– the quantifiable description of commitments regarding the QoS in the *SLS*, which includes the following information with reference to the SLS defined as part of the TEQUILA project [GOD 00]:

- *scope of the service*: this parameter describes the domain(s) affected by the delivery of the service,

- *description of streams*: this parameter indicates the streams to which the QoS must be applied,

- *conformity of traffic*: this part describes the parameters that the traffic must comply with in order to conform to a profile. If it does conform, the traffic will benefit from the guarantees given by the performance parameters (described in the following point).

The *parameters of conformity* to a profile notably include:

– the peak rate;

– the parameters describing a token bucket profile (speed and burst);

– the maximum size of data units;

– the minimum size of data units.

The *QoS parameters* (or *performance parameters*) have limits which the service provider commits to guaranteeing the client. Four types of guarantee are made for the traffic conforming to the given profile:

– a maximum latency between the input and output of the domain covered by the SLS;

– a maximum time variation (jitter) between the input and output of the domain;

– a maximum drop rate within the domain;

– a minimum bandwidth in the domain in question.

– the *treatment* of packets which *exceed* the profile;

– the *methods of measuring* the service with reports and the frequency of these reports;

– the *time spans* during which the service will be guaranteed;

– the *reliability of the service* which indicates the maximum time possible for the non-delivery of the service and the maximum repair time.

One of the challenges in providing real and automated end-to-end management of these agreements, and thus in ensuring a continuous service, is propagating these agreements over different technologies associated with the service providers of the domains involved. Two aspects, linked to traffic engineering, must be taken into account with regard to this propagation of SLAs:

– first, long-term agreements must be made between domains and the various stakeholders in communication, thus enabling the definition of limits to be respected in order to include all SLAs which may be developed. This long-term aspect creates conditions for the implementation of an end-to-end QoS;

– the second aspect to be taken into account is short-term dynamic SLAs (for instance, which may be required by a stream for one session). These SLAs corresponding to a regular requirement must be taken into account by all of the domains involved, which should adjust the resources accepted for corresponding services within the limits of the long-term agreements made between the domains and depending on the availability of resources at that given time.

It is in this second aspect, involving a high level of dynamic in the negotiation of SLAs, where propagation is the most critical. The objective is then to translate the parameters defined in the SLAs into parameters for the configuration of the communication network. This aspect must also be taken into account to enable an automated management of these agreements.

2.1.2. *Transport level*

At the transport level, two protocols are mainly used in the Internet in order to provide the features of this layer: the Transmission Control Protocol (TCP) and UDP (User Data Protocol) protocols. This section will examine the services which these two protocols offer. We will also discuss the Real-Time Transport Protocol (RTP) transport protocol used for applications with real-time requirements. Alternative transport protocols proposed by IETF which offer different services will also be presented.

2.1.2.1. *UDP: User Datagram Protocol*

The UDP provides a transport service in disconnected mode. Therefore, the UDP, as defined in [POS 80], enables the sending of application messages, or datagrams, by adding a minimum of protocol mechanisms to them.

The elements of information mainly transported by this protocol are:

– the transport service access points (TSAPs) (source and destination ports);

– the length of the UDP datagram;

– the UDP checksum conducted on the header of the datagram.

This transport protocol provides no sequence number, order control mechanism, reliability stream or congestion. Therefore, UDP offers a message-oriented service without connection which provides no guarantee on the order or the reliability of the data delivery. Any QoS control mechanism required by the application level must be implemented by the application or by protocols using the UDP service.

The advantage of this protocol lies in its simplicity and the speed of service it offers, given that no stream control, congestion control and connection management are carried out by UDP. The protocol is mainly used for applications which need to control their transmission rate or when there are real-time requirements, for example.

Nevertheless, it is rarely used without the addition of extra QoS control mechanisms (e.g. error control). These extra mechanisms are implemented by protocols such as RTP (described in section 2.1.2.2), or by the application which is able to handle some of these mechanisms itself.

2.1.2.2. *RTP: Real-time Transport Protocol*

The RTP was proposed in [SCH 03], as a support for time-restricted data communications. It follows the principles defined in the application level framing [CLA 90]. This protocol is mainly used with the Internet for multimedia applications. RTP is not really a transport protocol in the sense that it does not provide the services that this layer usually offers (above all, it does not specify the access point of the transport service). For this reason, it is generally used above a transport protocol, most commonly UDP.

RTP enables the transport of timestamps and sequence numbers which are useful for the applications with time (and QoS) restrictions in order to:

– detect and recover possible errors;

– reorder data;

– eliminate time-elapsed data;

– synchronize the various data streams in a session.

Moreover, RTP integrates other types of information regarding the identification of streams determining the type of media associated with the stream and the interdependencies between the data units of this stream.

2.1.2.3. *TCP: Transmission Control Protocol*

The second transport protocol, with the widest coverage over the Internet, is the connected mode (TCP), defined in [POS 81]. This protocol offers a fully reliable and ordered end-to-end service implementing mechanisms for controlling streams, errors and congestion.

This protocol was designed to adapt to all the underlying networks, whatever their reliability and capacity. In theory, it is capable of using the maximum capacity available along the pathway taken while ensuring an equal distribution of this capacity between all the TCP streams using the same resources. When it detects packets being dropped, due to the expiry of timers or acknowledgment messages coming from a receiver, it adapts the transmission rate by applying a congestion control mechanism based on a transmission window. Below we will examine the limits of these standard mechanisms applied to certain types of networks.

This protocol is used to transport application streams which need total reliability and order but which do not generally have high time restrictions. Indeed, the mechanisms implemented by TCP in order to control congestion and ensure the total reliability of the transmission of segments will lead to a high level of variability in the speed and delay of transmission for the various segments. This means that real-time applications will experience problems in respecting time restrictions if they use TCP.

Therefore, TCP is not ideal for transporting real-time multimedia applications, however, it is very well adapted for applications without high time restrictions (mail, ftp, web, etc.).

2.1.2.4. *Conclusion on TCP and UDP protocols from a QoS point of view*

We note that the two transport protocols which are almost exclusively used for the Internet offer entirely different services:

– the UDP offers an unreliable and unordered service without connection, which does not implement any mechanism for the control of stream congestion or errors. UDP has the advantage of being a light protocol which enables the application, at the risk of causing congestion in the network, to transmit without any restrictions on the state or type of underlying network. Regarding the QoS restrictions of application sessions, the use of this protocol does not offer a guarantee of the complete and ordered delivery of data to the application due to dropping or disorder which could be caused by the communication system. The parameters of *reliability* and *order* are, therefore, *not guaranteed* by this transport service;

– the second transport protocol, TCP, in connected mode, adapts to the state of the network and ensures a fully *reliable* and *ordered* service. Therefore, it offers these guarantees to application sessions. However, the mechanisms for controlling reliability, order, speed and congestion operated by TCP may have a negative effect on the QoS time restrictions of the streams transported by this protocol. Indeed, it does not guarantee a given transmission rate.

2.1.2.5. *Transport protocols emerging from IETF*

Emerging transport protocols at IETF offer alternative services to TCP and UDP, which dominate the Internet. Above all, they offer intermediate QoS levels between TCP and UDP. Two principle protocols (Stream Control Transmission Protocol) (SCTP) [STE 00] and Datagram Congestion Control Protocol (DCCP) [KOH 06] offer more diversified services than the current "all or nothing" option given by TCP and UDP:

– *SCTP* offers a multistream allowing application data to be divided into several streams of data which are then transmitted independently. The numbering of sequences is distinct for each different stream made of a multiple stream. However, within the stream, order restrictions are respected. As a result, if drops or disorders are detected for one of the streams, the delivery of data by other streams is not affected.

– *DCCP* offers an unreliable transport service which is regulated by a mechanism of congestion control for streams of datagrams. This protocol enables applications to use a simple unreliable UDP-type service, but adds a congestion control similar to that of the TCP (thus, reacting to phenomena of congestion which may occur in the network).

2.1.3. *Session and application levels*

2.1.3.1. *SDP: Session Description Protocol*

One of the most extensively used protocols on the Internet for describing sessions of multimedia application is the *SDP* [HAN 98], which is transported by session control protocols such as Session Initiation Protocol (SIP) [ROS 02a], Real Time Streaming Protocol (RTSP) [SCH 98], Session Announcement Protocol (SAP), Hyper Text Transfer Protocol (HTTP) or Simple Mail Transfer Protocol (SMTP).

We will examine this SDP, especially with regard to the information transported by this protocol, used by almost all of the distributed applications (proprietary session description protocols are also used by some applications).

SDP (as defined by the working group Multiparty Multimedia Session Control (MMUSIC) of the IETF) aims to describe the multimedia session in order to announce sessions, invite to a session or to start a multimedia session. However, it does not describe the way of exchanging SDP messages between applications. It is, therefore, the objective of the protocols cited above (SIP, etc.) to control the initiation and routing of data which will be used as a support for SDP.

SDP can announce multicast-type sessions, and also offer information necessary for initiating unicast-type sessions. The information contained in SDP messages describing the various components of the multimedia session are as followed:

– the name of the session and its object;

– the length of the session;

– the different media which make up the session with the following information for each of these media:

- the type of media (interactive audio, video, etc.),

- the media format (H.261 video, moving picture experts group (MPEG) video, GSM audio, etc.),

- the transport protocol used to transport the media (RTP/UDP/IP and H.320);

– the information for receiving these media (addresses and ports);

– the bandwidth necessary for the session (optional).

Therefore, it can be noted that the information on the types of media will be exchanged between applications. However, the real QoS level, necessary for the session, is not explicitly given since only the bandwidth of the session can be specified in an optional way, and the format of the media is also known.

From a QoS negotiation point of view, IETF standards feature a recommendation [ROS 02b], which enables a negotiation between two entities of media format to be used during the session, in the form of an offer/answer model. This negotiation may be limited to a comparison of media formats between the terminal entities which are negotiating. Otherwise, the decisions may reflect the restrictions of the communication system, but no negotiation scenario is developed in this sense in the request for comments (RFC). This possibility is mentioned in the event of a renegotiation of streams making up a session when there is a feedback regarding network congestion.

However, the information describing an SDP session significantly limits the possibility of negotiations and implementation of QoS in the communication system. Indeed, the parameters necessary to ascertain the precise requirements of the session stream (speed, jitter tolerated, etc.), are only partially exchanged in this session descriptor. The negotiation of the QoS and the possible configuration of the communication system, depending on the requirements of the higher layers, are thus reduced.

SDP was initially designed to announce and initiate sessions, not taking account of the possible negotiations in these sessions. This means that SDP is not suitable in a number of current application scenarios. Indeed, several new applications require an advanced description of the session and a prior negotiation depending on the capacities available (such as the type of codecs, their characteristics, the transport characteristics, etc.).

2.1.3.2. *SIP: Session Initiation Protocol*

The *SIP* [ROS 02a] uses text format messages to configure, manage and terminate multimedia session, independently from the transport protocol and the session to be established. These sessions can be unicast or multicast-type. The initialization messages of the session used by SIP transport session descriptions which enable participants to agree on the media formats used during the session. Usually, this media description is given using SDP (see section 2.1.3.1).

The SIP standard offers to create an infrastructure made up of network entities (proxys) by giving a unique SIP address to end users whatever their location in the network. The users will, therefore, be recorded by these proxys and start their session via them.

A simple session signaling example is given in Figure 2.7. SIP carries out traditional functions of a session control protocol by exchanging the following messages: INVITE, RINGING, OK, ACK and BYE.

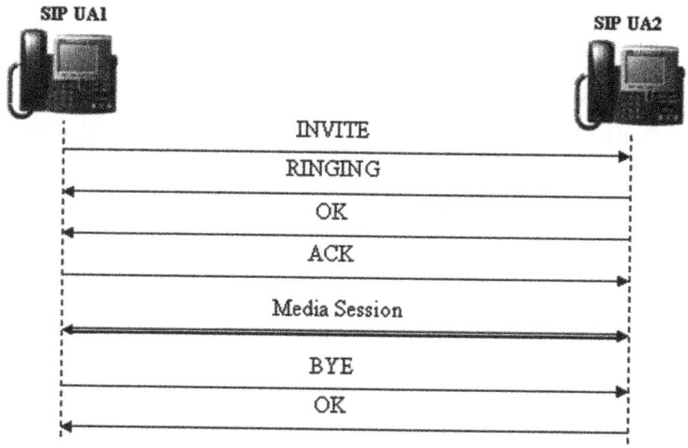

Figure 2.7. *Basic SIP session*

The SIP architecture defined helps to deliver the various services of recording, locating and redirecting sessions. The components of this architecture, as well as their respective roles, are detailed below:

– the user agent is the application of the end user. For instance, it may be a telephone or a videoconferencing terminal on IP, an audio or video server, or a pathway toward another protocol. It can be broken down into a client and a server part. The client part, known as the user agent client (UAC), sends SIP request, while the server part, known as the user agent server (UAS), receives the requests (or the answers of the distant client);

– the registrar allows users to identify themselves and register themselves in the "SIP domain" by communicating their session initiation protocol (SIP), uniform resource identifier (URI) and IP address;

– the location server accesses a database updated by the registrar, which then allows a correspondence between the SIP URI and IP address;

– the redirection server helps locate the address at which the interlocutor may be reached by offering a list of alternative SIP URIs;

– the proxy relays the SIP signaling without maintaining the states linked to this session (stateless mode) or conserving the states on the progress of the session (stateful mode) by modifying the first SIP message (INVITE) so as to force the following SIP messages in the session to be relayed by the proxy. This mechanism is at its most useful if linked up to invoicing services, for example.

An extension of the SIP protocol [CAM 02], which enables the transport and negotiation of SDP descriptions during the initialization of the session, offers mechanisms related to the implementation of QoS. Therefore, a phase of QoS reservation is planned in the standard during the initialization or the renegotiation of a session, as shown in Figure 2.8.

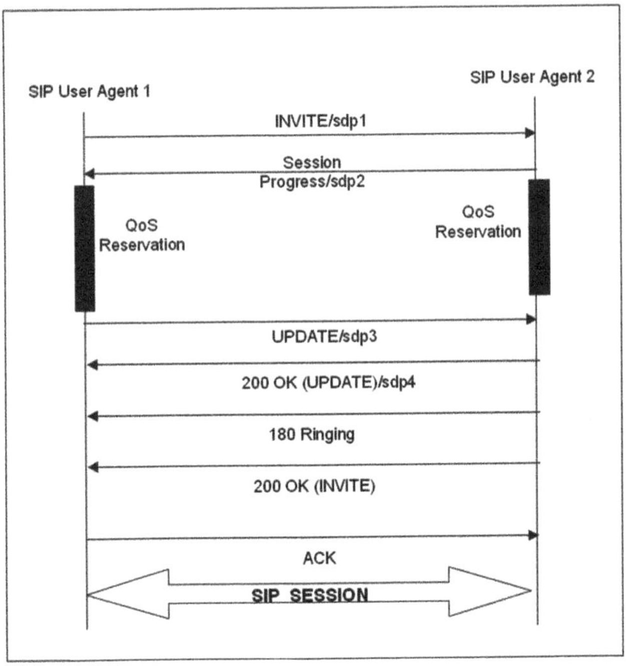

Figure 2.8. *Initialization of an SIP session integrating the quality of service reservation as per [CAM 02]*

It is then possible to use a combination of the negotiation of the SDP offer/answer session (see section 2.1.3.1) and this SIP mechanism in order to negotiate the media formats and the implementation of QoS in a session. However, these SIP extensions do not specify the content of the reservation phase of this QoS.

SIP currently appears to be the protocol which will enable the integration of voice/video/data services and the formal separation of service, control and transport layers, while offering standardized protocol and application interfaces between these layers. Its generic nature, openness, flexibility and the distributed nature of its architecture make it the protocol which is able to unify NGNs and act as a real support for the models of the most open stakeholders. This vision is confirmed above all by the adoption of SIP by 3rd generation partnership project (3GPP) as the call and invocation control protocol for application in the IP multimedia subsytem (IMS) [ETS 05a] of universal mobile telecommunications system (UMTS) networks.

2.1.4. *QoS signaling*

2.1.4.1. *NSIS*

The IETF group "Next Step in Signaling" (NSIS) aims to standardize a general IP signaling protocol with QoS signaling as the first case study. A number of concepts and general features of the protocol have been defined [HAN 05] and are presented in this section.

The working group has defined a dual-level signaling paradigm which is detailed below. The intention is on one hand to reuse the mechanisms of the RSVP protocol and on the other hand to simplify it while offering a more general signaling model. In order to follow an approach by operational blocks, this dual-level model separates the transport of the signaling from the application of the signaling. To do this, the protocol stack is separated into two layers: a generic lower layer and a specific higher layer for each signaling application. This enables the development of a signaling protocol which can provide signaling for various services or resources (such as network address translation (NAT), crossing firewalls and QoS resources).

NSIS transports information regarding streams of data along their network pathway and interacts with the nodes crossed. NSIS messages, therefore, cross the same nodes as data. The objective is to have a useable protocol over a number of Internet pathways, for various requirements, without necessarily having an end-to-end deployment.

In traditional signaling architecture, messages are received, processed and sent by signaling entities (SEs) which may be placed at nodes of the data pathway, or at the nodes which do not form part of the data pathway. Two types of signaling architecture are possible:

– distributed architecture, in which the signaling entities are in all the data pathway nodes (e.g. routers);

– centralized architecture, in which signaling is managed by a signaling entity. This architecture reduces the signaling load in the core nodes of the domain.

Detail of the model at the protocol suite layer

In order to form the modularity required by the NSIS demands, the dual-level protocol suite contains:

– the NSIS Transport Layer Protocol (NTLP), also known as general internet signaling transport (GIST) [SCH 06], which transmits signaling messages between the network entities. This process must be independent from the application of signaling;

– the NSIS Signaling Layer Protocol (NSLP), which includes the specific features of signaling applications. As can be seen in Figure 2.9, this dual-level protocol model enables the use of various signaling applications, such as QoS signaling [MAN 05] and NAT signaling and firewalls [STI 05].

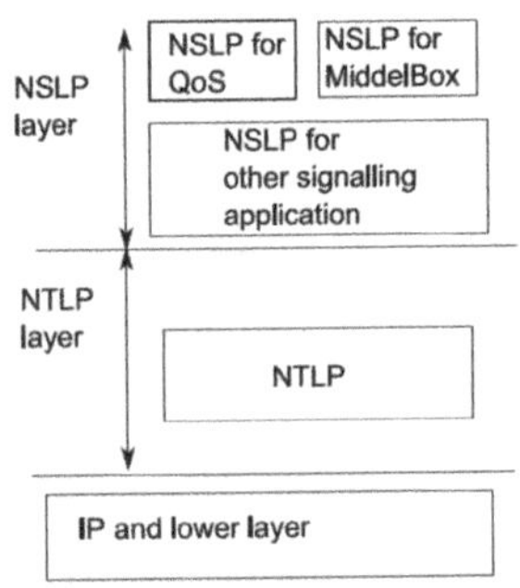

Figure 2.9. *Signaling protocol architecture*

The GIST layer transports signaling messages. GIST enables operations in connected or datagram modes (C-mode or D-mode). When a message is ready to be transmitted, it is sent to the GIST layer, with information regarding the corresponding stream. It is thus the GIST layer which transfers it to the next network element (NE) along the pathway. In the example given in Figure 2.10, the first NE of the network pathway includes the transport layers NTLP and NSLP 1. NSLP 1 generates a signaling message and sends it to the local NTLP layer so that it can be transmitted to the destination stream. In the second and third network entity (NE), the NSLP layer does not exist or is not the same as in the first NE. The message is then transmitted without being processed by the NTLP transport layer. Finally, the signaling message arrives in the last network entity where the NSLP signaling application is the same as in the first network entity. The NTLP transport layer thus recovers the message and transmits it to the NSLP layer, which applies the corresponding processing.

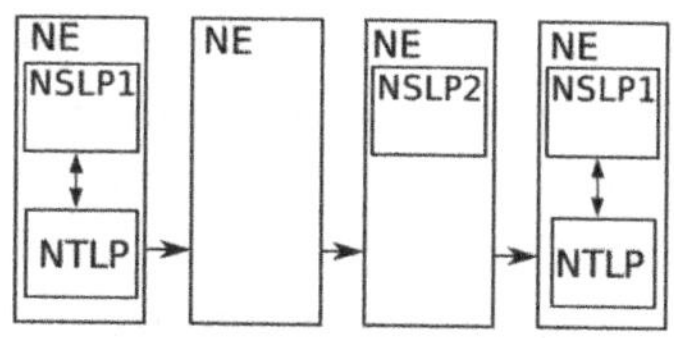

Figure 2.10. *Signaling* via *heterogeneous NSLP applications*

Signaling for quality of service

In the case of QoS signaling, the basic NSIS concepts apply and the direction of the signaling process is added, so that one of the extremities of the signaling stream takes the responsibility of reserving resources.

A set of extra entities is, therefore, defined in this framework:

– the QoS NSIS initiator (QNI): the signaling entity which makes resource requests, generally triggered by an application request;

– the QoS NSIS responder (QNR): the signaling entity which is a terminal signaling element and can potentially interact with applications;

– the QoS NSIS forwarder (QNF): the signaling entity between the QNI and QNR which propagates the NSIS signaling in the network.

Each of these entities interacts with a resource management function (RMF), which effectively allocates network resources (buffers in routers, bandwidth, regulation and traffic shaping).

The QoS signaling layer (QoS NSLP) includes a set of messages to reserve resources along the signaling pathway. The principle messages are as follows:

– the "Request" message creates a reservation for a stream;

– the "Modify" message modifies an existing reservation;

– the "Release" message drops an existing reservation;

– the “Accept/Reject” message confirms or rejects a QoS reservation request;

– the “Notify” message reports an event detected in the network;

– the “Refresh” message is linked to the state management mechanism.

Messages can be sent or received at QNF, QNI and QNR nodes. An example can be a QNF located at the borders of the network which configures the resources for a stream crossing the domain. It is not specified whether the answering node can free or modify a reservation during or after the configuration. This is dependent on authorizations and the characteristics of the type of resource in question. The “Refresh” message, available in QoS NSLP, does not modify the reservation but enables an extension of its lifespan and provides a state management mechanism.

The traditional NSIS signaling process is shown in Figure 2.11, for a resource reservation which follows the data path.

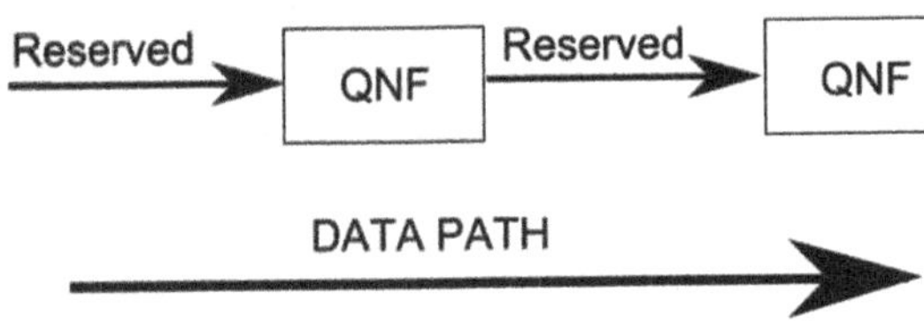

Figure 2.11. *Traditional NSIS signaling process*

The NSIS signaling layer (QoS NSLP) is not itself involved in the specific techniques for management or allocation of resources. However, the definition of the NSLP layer for the reservation of QoS involves a notion of admission control. For an NSLP QoS layer, successful signaling corresponds to the capacity of reserving part of the total resources available in the network. The RMF is responsible for the functions of supply, surveillance and insurance. In this model, the NSLP is a client of the RMF. It should be noted that the RMF can in turn be the client of another element of the NSLP in order to effectively implement the reservation of resources in the network. This aspect will be important for signaling at several levels: for instance, for two levels, a high level of signaling taking the responsibility of Internet QoS, possible from end-to-end, and a lower signaling layer taking charge of the more specific intradomain QoS, between the borders of the network in question.

2.1.4.2. *COPS*

IETF defines the common open policy service (COPS) [BOY 00] as the standard transaction protocol in policy based networks (PBNs). This protocol can also be used alongside other signaling protocols in order to implement the QoS (use of RSVP or NSIS combined with COPS).

The architecture of policy control [YAV 00] is a new approach for automatically deploying a policy within an administrative domain. A policy consists of a set of rules for administrating, managing and controlling the access to network resources [WES 01]. Each policy corresponds to a set of conditions associated with a series of actions to be carried out if the conditions are verified. In order to reuse the policies on different management domains and to deploy them easily throughout the network, a policy core information model (PCIM) [MOO 01] has been defined by the working group of the IETF policy. It is explained in detail below.

Three concepts are inherent in the general policy definition when they are applied to a network environment:

– the abstraction of policy rules: a policy may exist at various levels ranging from business objectives to configuration parameters depending on the network architecture. Translations between these different levels of abstraction may need information other than policy rules, such as network configuration possibilities and the terminal system in question;

– policy defining an action: this type of policy defines what must be done in order to ensure compliance with a policy rule, when conditions defined by this rule are unified. Policies of this type define the executions of one or several operations which may, for example, have an effect on and/or configure the traffic and resources of a communication network;

– policy defining a condition: this type of policy corresponds to the representation of the necessary state to be in and/or the necessary prerequisites determining whether the actions associated with the policy rule in question must be executed.

The framework proposed in PCIM uses these three concepts and presents an object-oriented information model in which policies are applied by using a set of conditions which need to be verified before executing a set of corresponding actions:

– the set of conditions associated with the policy rules specifies the applicability of the rule. The set of conditions can be expressed by conjunctive or disjunctive normal forms. These conditions may be simple or composed;

– if the set of conditions associated with a policy rule is evaluated as true then a set of actions is applied to the object. The object can stay in the same state or can alter toward a new state.

In order to facilitate a hierarchical approach between the policy rules, *groups* of policy rules, bringing together a set of rules, are introduced. These groups can be combined in order to represent hierarchies in policies.

A notion of priority is also introduced in this framework. Indeed, the execution of policy rules can be prioritized. The advantage of this priority-based approach is that it expresses an overall policy which has general rules and some specific exceptions.

The concept of *role* is also defined in this framework. It helps to attach a set of policy rules to entities which have the same role in the communication network.

Figure 2.12 summarizes the various concepts proposed by the PCIM approach in the form of a flow chart.

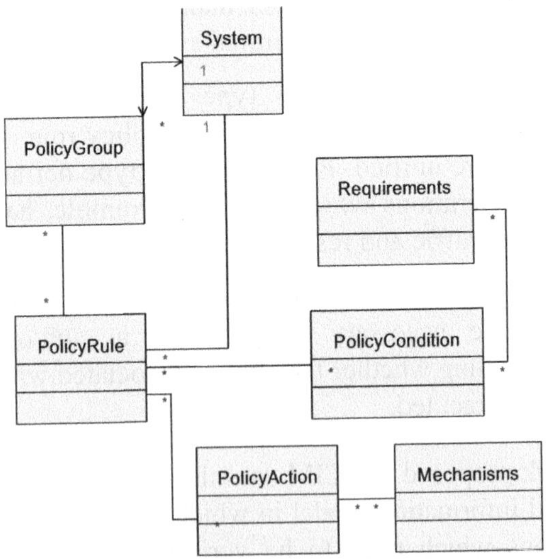

Figure 2.12. *Flow chart of PCIM architecture*

Regarding the implementation of QoS, a hierarchical breakdown of this model is, therefore, proposed in order to abstract it from both: the application domain (e.g. QoS and security), the technology used (e.g. DiffServ and IntServ) and the manufacturing specifics of the equipment (e.g. Cisco and Lucent). In the case of a QoS implementation policy, a DiffServ network operator, for instance, deduces from these different SLAs of policy rules abstracted, which are translated into generic configuration commands for routers, and then a sequence of specific commands for the router in question.

This architecture uses various components as shown in Figure 2.13.

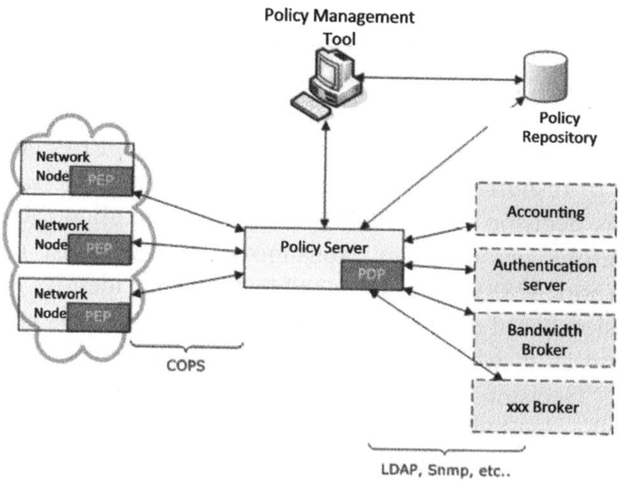

Figure 2.13. *Policy control architecture*

As shown in Figure 2.13, the COPS approach in the PCIM framework uses the concept of role by defining two main roles: the "Policy Decision Point" (PDP) and "Policy Enforcement Point" (PEP); and a communication protocol between the two:

– PDP is a logical entity which takes decisions for itself and for all network elements expressing a demand [YAV 00]. A rule server includes a PDP;

– PEP is also a logical entity which applies or carries out decisions and translates the high-level rules into configuration instructions of various devices;

– COPS is a simple request/answer protocol based on TCP which is used to exchange information regarding policies between a PDP and its clients (i.e. PEPs) [BOY 00].

Moreover, a set of components can be defined as part of the application for the implementation of the QoS:

– a policy repository stores the resource management rules;

– a policy management tool adds new rules and removes others;

– a BB centralizes the state of bandwidth resources in the network.

There are two management models in policy-based architectures:

– the *outsourcing* model [HER 00], in which the PEP requests the PDP respond to an event on the basis of rules which the PEP does not have. This model is usually associated with end-to-end signaling protocols (RSVP-type). For instance, events signaled on the PEP entity for the "reservation of resources" may require an accept/reject decision. The PEP delegates this decision to the PDP by sending an explicit "Request"-type message. Before any data are exchanged, the PEP must open a TCP connection (port 3288) to its responsible PDP and identify it. The PDP takes a decision based on the state of the network resources and the policy of the domain as to whether this request can be accepted. It answers the PEP with a "Decision"-type message which stipulates the configuration which may be implemented. When the PEP has established the new configuration, it must inform the PDP. The PDP can then modify or update the configuration information with new decision-based messages. For each decision, the PEP removes the named configuration and sends a confirmation to the PDP.

– The *provisioning* model [BOY 00], where the PDP prepares the configuration of the equipment using new network use rules (implementation of a new SLA, change of management rules, time/date, expiry of quota, etc.) The PDP sends the configuration information to the PEP without an explicit demand, and the PEP will apply these policies in the necessary time without further notification. This model is above all used to control the use of the network with minimal signaling (e.g. DiffServ). In this model, the resources are generally preallocated which allows the exchanges between the PDP and PEP to be minimized.

The COPS protocol thus transports objects between the PDP and PEPs in a transparent way. Each object is made up of class instances whose formats and attributes are defined by a tree structure of data named policy information base (PIB). For this reason, the protocol can be easily extended and offers (via the PIB) reusable generic classes for different types of policy. Therefore, the DiffServ PIB [CHA 03] defines the capacity tables which the PEP uses to inform the PDP of its capacities and parametric syntax of the various processing functions of a DiffServ packet.

2.2. ITU-NGN approach

2.2.1. *Principles*

The communication architectures are currently evolving toward a general infrastructure based on IP: NGNs. This evolution can be explained by the IP's high integrating power which offers a routing method of independent data: on one hand, of the type of underlying network technologies, and on the other hand, of the type of data transported. The objective of telecommunications operators is, therefore, to produce, via these NGNs, support for multiple services (telephone, television and Internet services) within a single infrastructure which may draw some heterogeneity from network technologies.

Standardization work within ITU is underway regarding the convergence of these networks, and therefore the evolution of traditional telecommunications networks toward complete IP networks. Indeed, the ITU working group on NGNs is currently standardizing new-generation generic architecture which integrates heterogeneous networks, especially for wireless access networks, thus enabling the convergence of voice and data networks.

Figure 2.14 shows the various types of access networks and associated technologies which the ITU aims to integrate in its unified architecture.

The classification presented shows the degree of mobility of various access networks and the data-rate potentially provided by these networks according to the type of mobility in question.

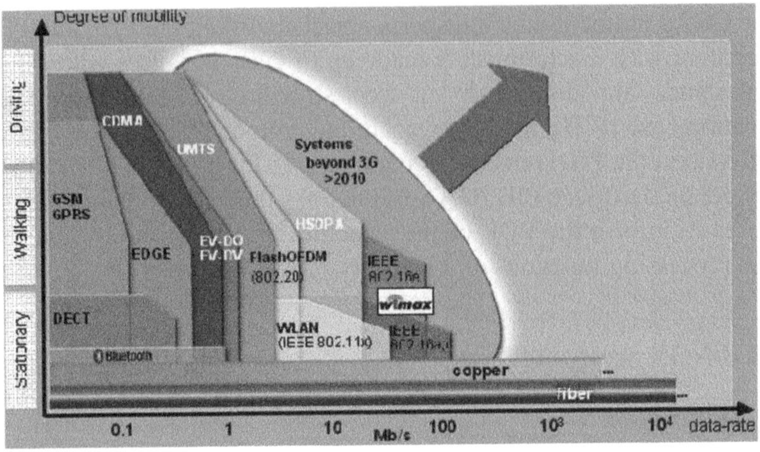

Figure 2.14. *Various access networks to be integrated into NGNs by ITU (copyright ITU)*

In order to provide a unified network to integrate existing telecommunications infrastructures with new ones in a single, common, flexible and evolving infrastructure, requirements were clearly identified by bodies defining NGNs, especially ITU-T (which is offering a combination of these approaches). These requirements can be summarized into five points:

– an evolution of data transfer toward the packet method;

– a single and pooled core network for all types of access networks and services;

– open and standardized interfaces between each layer in order to achieve the independence of the service from the network;

– a support of multiple access technologies;

– a multiple-terminal support (modular, multimode, multimedia and adaptive).

After the evolution of the core network toward "full IP", the most important notion remains the break down into separate functional plans by open interfaces which ensure both the scaling-up and the flexibility by offering an interconnection and integration facility for new services.

At this stage, there is not yet a single definition for the notion of NGN. However, ITU-T is offering characteristics and a generic architecture of multinetwork, multiservice, multiprotocol and multiterminal NGNs. Indeed, an architecture using the end-to-end IP protocol and integrating heterogeneous networks is proposed based on decoupling of services offered by the network, of the real transport of data and of the management of this transport.

This generic architecture, shown in Figure 2.15, is therefore broken down into two principle levels which correspond to a service layer and a data transport layer.

One of the objectives of convergence toward a unified network is to be able to ensure end-to-end QoS guarantees in a heterogeneous network using a single transport management. To do so, the generic architecture of the NGNs integrates a set of features relating to the implementation of the end-to-end QoS.

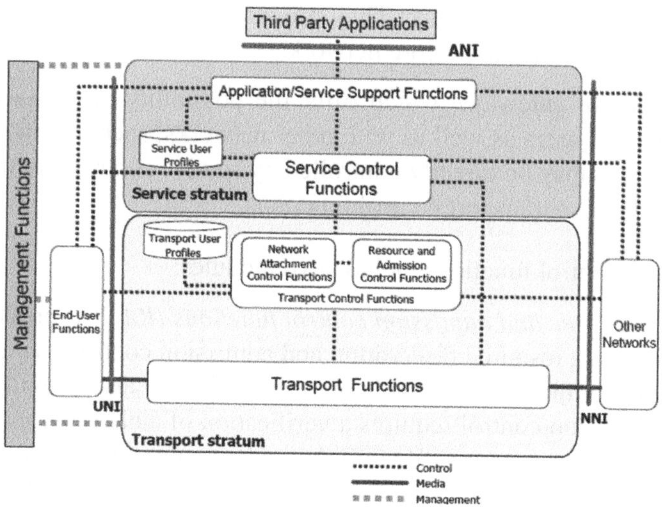

Figure 2.15. *General architecture of NGNs according to ITU*

The various functions offered by this generic architecture of NGNs defined by ITU-T are detailed below.

2.2.2. *Transport stratum*

The *transport functions* provide the connectivity to the set of components and functions in the NGN. These functions provide a support not only for the transfer of data, but also for the transfer of control and management information.

The transport functions, therefore, include:

– access network functions, which correspond with access functions of users of the core networks, and the implementation of QoS mechanisms to the user traffic, such as the management of buffers, packet filtering, traffic classification, marking, shaping and regulation of traffic;

– the functions of border elements, which are used to process traffic coming from various access networks and aggregated upon entering the core transport network. These are support functions for the QoS and controlling traffic;

– core network functions, responsible for ensuring the transport of information in the core network. They provide the means for differentiating the quality of transport in the core network;

– functions of gateways, which offer the possibility of interacting with functions of end users as well as with other networks, such as Internet users. These functions may be directly controlled by service control functions or by the intermediary of transport control functions.

Transport control functions include two modules:

– First, *resource and admission control functions (RACFs)* provide a QoS control, including resource reservation and admission control, via access and core networks. Other functions relating to security are also provided by this module. Admission control requires a verification of authorizations based on a set of parameters: user profiles, SLAs, policy rules specific to operators, service priorities and the availability of resources in access and core networks.

At the core of the NGN architecture defined by ITU-T, the RACF plays a marshalling role for negotiating and allocating resources between service control functions and transport functions. The RACF interacts with these functions for applications based on sessions (e.g. SIP calls) and for applications which are not session-oriented (e.g. video streaming), but which require resource control, for reasons of security or QoS deployment. It thus interacts with transport functions to control in the transport layers one or several of the following functions: the classification of traffic, the marking, shaping and regulation of traffic, the establishment of priorities, the reservation and allocation of resources, the translation of addresses and ports and the crossing of firewalls.

The RACF also interacts with *network attachment control functions (NACFs)* to verify user profiles and SLAs which they have set up.

– Subsequently, NACFs include the registering of users, the authentication, authorizations and parameters of configurations.

Indeed, the NACFs provide the registration features at access level and the initiation of functions enabling users to access NGN services. These functions allow the identification and authentication at the network level. It also informs users of the point of contact for the support of NGN applications and services. The detailed features provided by the NACF are as followed:

– the dynamic allocation of IP addresses and other parameters for the configuration of users' equipment;

– authentication at IP level (and potentially at other levels);

– network access authorization, based on user profiles;

– the configuration of access networks, also based on user profiles;

– localization management at IP level.

Transport user profiles functions are databases where user information and control data are pooled into a single "user profile" in a transport layer. This database can be distributed to a number of places in the NGN.

2.2.3. *Service stratum*

The service level defines three main modules:

– *Service control functions* include the control, registration and authentication of session at the service level. They may also include functions for controlling media resources, which are specialized resources and gateways at the service signaling level.

– *Application/service support functions* include features such as gateways, registration, authentication and authorization functions at the application level. These functions work alongside service control functions to give users and distributed application providers the added value which they need. The communication interface between these functions and the user is the *user-to-network interface* (UNI), while the interface between these functions and the application is the *application-to-network interface* (ANI).

– *Service user profile functions* represent the pooling of user information and other control data into a single user profile stored in a database.

2.2.4. *Management plan*

The management aspect is a fundamental element of NGN operations. *Management functions* offer the possibility of managing the NGN so that it can provide services of the desired quality, security and reliability. These functions are allocated throughout each functional entity and they interact with the management modules of the network entities, the general management of the network and the functional entities of the service management. Management functions are applied to the service and transport strata in the NGN. They cover the following functionalities for these two levels:

– error management;

– configuration management;

– compatibility management;

– performance management;

– security management.

2.3. Conclusion

This chapter presented a state-of-the-art on the techniques of management and implementation of QoS in terrestrial networks. Two complementary approaches were discussed – for the Internet, standardized by IETF and for NGNs, standardized by ITU-T.

The different approaches in the definition of QoS are linked to the issues relating to the various types of standardization. Hence, IETF closely examines and defines the protocols which may be used in a generic way in order to provide a QoS service in networks, while ITU focuses more on the architecture of the QoS, using mechanisms defined elsewhere.

This is also due to the composition of these groups, since, there are mainly Internet constructors and stakeholders (especially, providers), while telecom operators have a significant influence at ITU. It is also possible to observe some stakeholders looking to implement controls, for example for IMS architectures, to added value services.

QoS can, therefore, be ensured in an open environment with a distributed approach (which is more in line with the IETF vision), or with a more centralized approach by the operators themselves (ITU approach).

There are a number of links between these two architectures, as discussed in the chapters related to the integration of satellite networks in next-generation terrestrial networks.

Chapter 3 examines the implementation of QoS in satellite networks, and more specifically in DVB-S2/RCS networks, which now offer the most possibilities in this domain.

3

Quality of Service in DVB-S/RCS Satellite Networks

Although geostationary satellites have been used for broadcasting services for some time, significant efforts have been made by the "satellite community" to make it into a supplementary technology for terrestrial infrastructure (for the support of broadband interactive service in areas not covered by terrestrial networks). These advances aim to overcome the major obstacles to satellite networks which are:

– the high costs of equipment and satellites;

– latency time (a 500 ms propagation delay which cannot be compressed as part of direct communication);

– weak and costly bandwidth.

One of the first advances is the adoption and specification, by the community, of interfaces and open standards in order to significantly reduce equipment costs and offer reasonable service prices to users, promoting competition between operators, content providers and manufacturers. This trend has been confirmed by the success of the DVB-S standard, which is currently the favored broadcasting format throughout Europe and by the spread of receiver terminals. The standardization of a digital video broadcast return channel over satellite (DVB-RCS) [ETS 09a][1], as well as efforts to achieve the interoperability of equipment made by the SatLabs group align directly with this trend.

---

1 These ETSI standards can be consulted for free on the ETSI website (http://www.etsi.org/WebSite/Standards/Standard.aspx). Search by document number.

By providing satellite terminals with a satellite return channel, DVB-RCS clearly surpasses satellite systems with a terrestrial return channel, making them a support for interactive access to broadband multimedia services.

Numerous recent technological advances have further increased the potential of bi-directional satellite systems as access networks. Therefore, the use of Ka band (27 to 40 GHz), previously highly dependent on meteorological conditions, allows satellite networks not only to benefit from wider channels (compared to the hugely congested Ku band), but also to reduce the size of user terminals and antennae. This was made possible due to the reduction in beam coverage.

The use of multibeam coverage in a transparent or regenerative framework leads to a subsequent gains in capacity, due to frequency being reused when possible. Coupled with a regenerative payload, multibeam coverage allows for greater interconnection flexibility for an optimal use of resources.

Finally, the DVB-S2 standard [ETS 09b], a successor of DVB-S, offers coding and broadcasting techniques which are more adapted to point-to-point services required by new multimedia applications with an improved use of capacities of a transponder ranging between 100 and 200%. These evolutions indicate a predominant place for satellite networks in the future of access technologies.

Before examining the details of the DVB-S and DVB-RCS standards on which traditional bi-directional satellite architectures rely, we will take a closer look at the main components and features of these architectures.

3.1. Bi-directional satellite access systems

3.1.1. *Overview*

Broadband multimedia satellite systems, whatever the implementation in question, can be broken down into three distinct segments:

– the user segment which consists of a return channel satellite terminal (RCST or ST) integrating satellite access features and ensuring interconnection between user equipment and satellite network;

– the spatial segment which is made up of one or several *satellites*;

– the operator segment made up of one or several *network control centers* (NCC) and *gateways* (GW) interconnecting with terrestrial core networks.

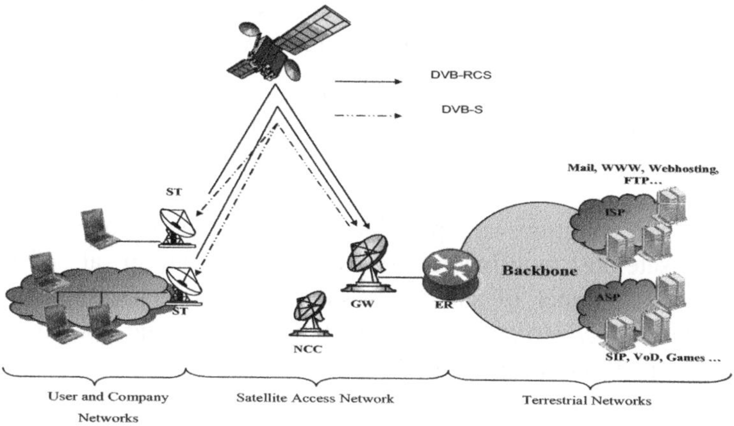

Figure 3.1. *Basic bi-directional satellite access infrastructure*

Figure 3.1 shows a traditional transparent star-shape satellite access network and its integration within an infrastructure such as the Internet.

The GW centralizes all the traffic in the satellite network. All the connections from the STs are directed toward the GW, hence the star-shaped structure. The GW emits in DVB-S on the forward channel (GW → ST) and receives in DVB-RCS on the return channel (ST→GW). Thus, the GW concentrates on the forward channel the traffic coming from internet service providers (ISPs) and has two administration modules for managing satellite resources, the network management center (NMC) and the NCC. These entities are responsible for the administration and control of the access of satellite resources by STs. Here, these entities are included in the GW. In a regenerative configuration, they are decoupled from the GW and unified within a single entity. A satellite network hub is generally at least composed of the three entities: GW, NCC and NMC.

The STs act as access routers in the satellite segment and concentrate the user traffic into the return channel. They emit in DVB-RCS and receive in DVB-S. They interact with administration modules of the GW in order to obtain their configuration parameters and their satellite resources.

3.1.1.1. *Access connectivity and connectivity mesh*

Two types of interaction are involved in these networks:

– *access connectivity* (GW-ST) which allows users connected to an ST to access the Internet (or other networks) via the GW;

– *connectivity mesh* (ST-ST) allows users connected to STs belonging to the same satellite network to communicate, either via the GW, or directly from ST to ST via the satellite.

Various configurations are possible to support these connections. The broadband satellite multimedia (BSM) working group at ETSI thus distinguishes a number of satellite system families [ETS 05] combining *mesh-* or *star-shaped* structures as defined in [ETS 01], and a satellite with a *transparent* or *regenerative* payload.

3.1.1.2. *Star and mesh topologies*

The mesh architecture supports direct communication between users connected to STs and belonging to the same satellite network, which is generally substantiated by symmetrical exchanges. In the star architecture, exchanges are all made via a central station, the GW, and are therefore asymmetrical in nature with higher data-transfer rates from the GW to the STs. However, this star architecture can also support mesh connectivity by establishing two indirect links between STs via the GW.

3.1.1.3. *Transparent or regenerative satellites*

Internet access can be made via two channels:

– the *forward channel* coming from a SP or Internet access (and therefore from the GW) to the user who takes some of the broadcasting capacities of the satellite to several STs;

– the *return channel*, from STs to the GW, on which the satellite presents a collection capacity by receiving information coming from various STs.

The management of resource sharing is not the same. Resources in the forward channel are monopolized by the GW, while resources in the return channel are shared between all the STs. Thus, ETSI has specified two standards, one standard for the forward channel which is DVB-S, and the other standard for the return channel, the DVB-RCS. The value of a transparent or regenerative then becomes clear since, depending on the payload of the satellite, the conversion of DVB-RCS signals into DVB-S signals takes place either on the ground in the GW for a transparent payload, or in the satellite for a regenerative payload. Both cases are shown in Figure 3.2.

Therefore, with a transparent satellite, the inter-ST communication requires a "double hop" satellite while a regenerative satellite only requires a "single hop".

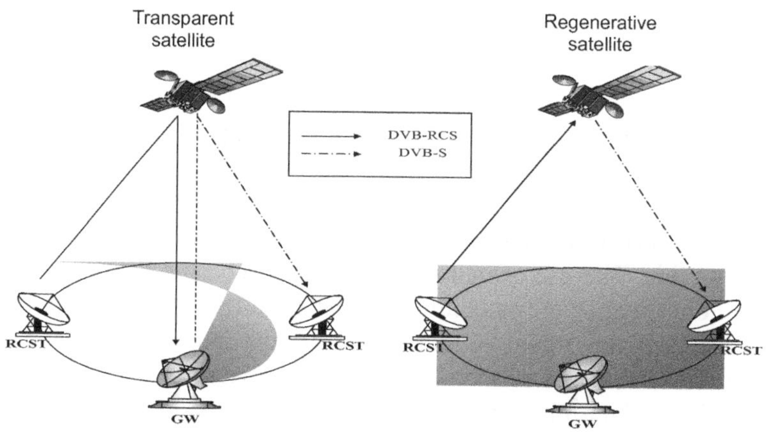

Figure 3.2. *Inter-ST communication with transparent and regenerative satellites*

The DVB-RCS standard specifies that *uplinks* (toward the satellite) and *downlinks* (from the satellite) must use different frequencies.

Thus, in *transparent mode*, the transponder[2] of a satellite acts as a simple repeater which receives a pre-determined channel via an uplink, converts its frequency and returns the channel to the downlink. The amplitude and frequency of the carrier are modified, but the modulation and the spectrum shape remain unchanged.

2 Transponder: in-built repeater in the satellite which receives the signals of an uplink on a pre-determined channel, converts it to the frequency of the downlink and returns it to the downlink. Transponders are associated with one or several emitter antennae which, through their form and orientation, define the coverage area of the beam emitted.

In *regenerative mode*, the transponder has the same features, as well as an on-board processor, which includes the digital processing of the signal enabling the demultiplexing/remultiplexing of the streams within the satellite. This is known as on-board processing (OBP). It is also common to see the terms regenerative payload or regenerative satellite used. As well as reducing the propagation delay between STs, this technology, combined with multibeam use, helps to optimize the management of satellite resources.

Although the interactivity of inter-ST communication has increased, it has done so at the expense of the simplicity and cost of the satellite.

3.1.1.4. *Single beam and multi-spot satellites*

Each channel can be defined by its emission and reception zone. These different zones offer a processing similar to a group of terminal satellites which share the same service. A spot thus delineates a coverage zone generally associated with one or more transponders. Most traditional satellites are usually single-beam, which means that coverage zones are particularly wide. The use of multiple low aperture beams helps to separate the overall coverage of the satellite into smaller zones. Since the size of the spots is reduced, the signal is more concentrated and its gain is higher. Therefore, for an antenna of an equivalent size, the terminal benefits from a more significant data-transfer rate. Moreover, the support of several spots enables a reuse of frequencies from one spot to another, and a non-negligible gain regarding point-to-point messages. This is because instead of being transmitted to a particularly wide spot, they are directly oriented onto the corresponding spot, thus saving the resources of the other spots. The interconnection between spots in a regenerative framework is all the more interesting and effective given that it enables a combination, in various target download spots, of different upload streams coming from different upload spots.

These restricted multi-beam technologies have existed for several years, in the form of IPStar or Astra 1H (made up of 32 transponders and 8 Ka band beams). The SkyplexNet/Hot Bird 6 systems as well as Amerhis [AME 04] support regenerative capacities coupled with multi-beam techniques. Figure 3.3 shows a bi-directional satellite system integrating multi-beam techniques and a regenerative payload.

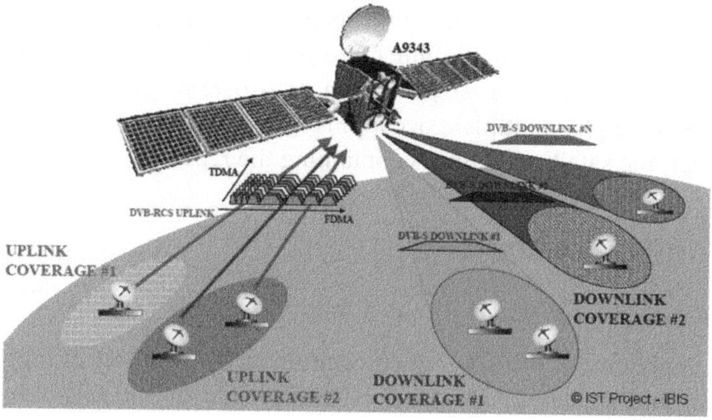

Figure 3.3. *Regenerative multi-spots bi-directional satellite*

Having highlighted the components and fundamental concepts of bi-directional satellite systems, this chapter will examine the details of the DVB-S and DVB-RCS standards which define the mechanisms on physical, data link and network level. To better understand the difficulties of integrating satellite access networks into NGNs, we will focus on the interactions with IPs envisaged in each standard.

3.2. The DVB-S standard and the IP support

Initiated in 1993, the European and international DVB project published a set of specifications at the end of the 1990s based on a standard for transmitting video and sound in digital form in a compressed format[3] developed by the MPEG group. This format is the MPEG-2 [ISO 00a] which defines both a compression standard and multiplexing methods. DVB was broken down into as many specifications as transmission media, since the three main standards were specified for cable (DVB-C), satellite (DVB-S) and terrestrial (DVB-T).

DVB-S thus defines the transmission mode, over a mono-directional satellite, of audio and video data toward terminals fitted with a DVB-S card. Data other than multimedia streams are also supported. DVB systems transporting IP packets are known as DVB Data Broadcasting systems

3 Range of compression ranging from 52:1 for the MPEG1 in 1991 to 200:1 for the MPEG-2.

[ETS 04]. Compared to the bi-directional architecture discussed in section 3.1.1, the traditional architecture of a DVB-S system designed for televisual services uses a single beam transparent satellite and a GW generating a single multiplex occupying the complete bandwidth of the transponder, the satellite terminals containing the receiver.

3.2.1. *The DVB-S standard*

The success of this standard comes from:

– the simplicity of the GW which centralizes the data (set of programs to broadcast) and signaling. This data is multiplexed by the GW on a single carrier occupying all the bandwidth offered by the transponder. The GW emits a steady flow using padding bits where necessary;

– the speed and ease of installing satellite terminals;

– the evolution of capacities offered by the general programming of MPEG2-TS containers (in order to include the video and audio streams and all types of private data).

3.2.1.1. *MPEG2-TS transport streams*

Whatever the type of data emitted by the gateway toward satellite terminals (sound, video or data), they are encapsulated in MPEG2-TS packets of a fixed size (188 bytes), whose sequence creates a MPEG2-TS transport stream. This is also a case of multiplexing since it groups several data channels within a single communication channel. A number of methods, as shown in section 3.2.1.2, are available in order to encapsulate various types of data. Here we will examine the broadcasting of video and audio streams which made DVB-S successful in the framework of digital television. Figure 3.4 illustrates the elements involved in the multiplexing of different programs.

Let us imagine a satellite package offering various television programs to the subscribers. Each program is made up of a video source and one or more audio and data sources (e.g. for subtitles) synchronized by the Program Clock Reference. Each source is compressed separately and gives a raw signal following compression known as the elementary stream (ES). These elementary continuous audio and video streams are then encapsulated in packets (PES: packet elementary stream) using a common reference base. Each packet thus contains a timestamp and a stream identifier (audio or video). The size of this packet depends on the type of application and can reach several hundred kilobytes. The set of programs of the satellite package, broken down into as many PESs as sources, is finally multiplexed within the MPEG2-TS transport stream. The packets constituting this multiplex are MPEG2 containers independent of the type of traffic transported (native MPEG2 source, or IP). The encapsulation method implemented is data streaming. According to this principle, PESs are fragmented and inserted into MPEG2-TS packets. Since PES packets must always coincide with the first byte of the payload of a MPEG2-TS packet, padding bits are often necessary.

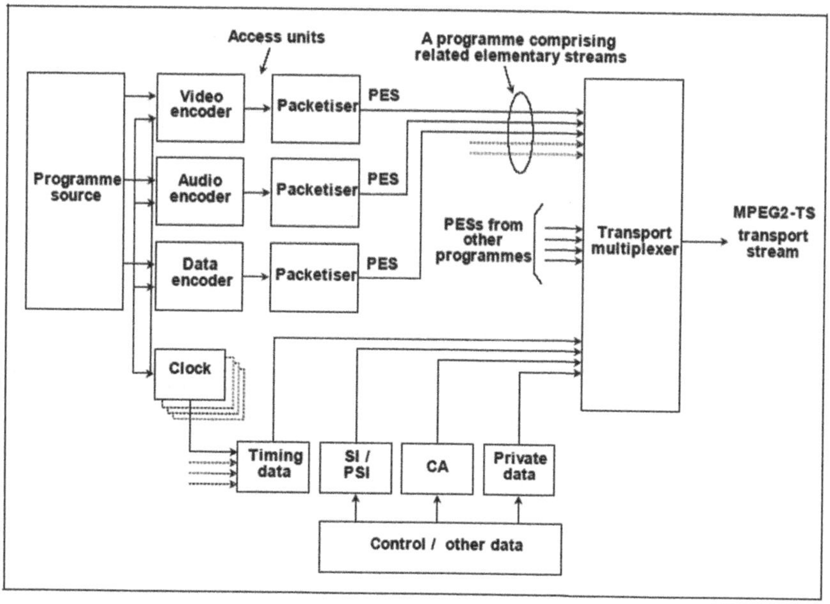

Figure 3.4. *MPEG2-TS multiplexing*

MPEG2-TS packet

Whatever the packaging method used, the format of the MPEG2-TS header (at least 4 bytes) remains the same (see Figure 3.5). The fields which we are most interested in are:

– *PUSI* (1 bit), which indicates whether the start of a piece of data is encapsulated in the packet (for data streaming, this corresponds to the start of a PES);

– *PID* (13 bits), which helps to identify a logical channel (ES) within a multiplex (some are dedicated to signaling);

– *adaptation field* (2 bits) in which the presence of private data can be specified.

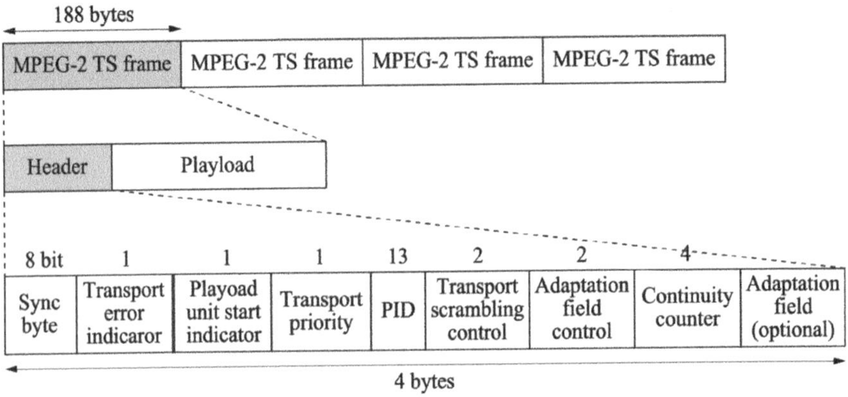

Figure 3.5. *Format of a MPEG2-TS packet*

3.2.1.2. *DVB-S signaling: PSI/SI tables*

A DVB-S receiver therefore selects an ES within a multiplex depending on the PID associated with the header of each MPEG2-TS packet. Since a program is made up of several ESs, the receiver must be able to identify the different services available from the PIDs connected to the same program. To do this, the receiver uses PSI/SI tables (program and service information tables).

In this section we will simply give a brief presentation of PSI/SI tables using a short example and advise the reader to refer to the following standards for more information: the ISO MPEG2 standard [ISO 00b] for obligatory PSI tables, the DVB-SI standard [ETS 12a] for optional SI tables.

In order to illustrate the use of these tables, we will provide a simple example. On a multi-transponder satellite, the DVB-S terminal of the user settles in a reference channel allowing access to the network information table (NIT) which groups the set of transponders and services made accessible to the subscriber by an operator. The operator chooses the multiplex containing the services which they are interested in. The terminal settles on a corresponding transponder using information from the NIT. The decoder reads the program association table (PAT) which informs it of the services available. Each service has a program map table (PMT) associated with it which references the characteristics of the service and its components. By switching from one program to another, the terminal reads the associated PMT which gives it access to the PIDs of the elementary audio and video streams making up the program.

These tables are multiplexed within the transport streams and are usually emitted periodically by the control entity or the SPs. ETSI defined a separate structure for each table, based on a common structure with DSM-CC sections (digital storage media – command and control) [ISO 00b]. A private section is transmitted via several MPEG2-TS packets using the same logical channel within the TS multiplex.

3.2.2. *Access method*

As we have seen previously, traditional DVB-S architectures generally only consist of a single GW which only emits a multiplex occupying the total bandwidth of the targeted transponder. Consequently, no access method is specified in the DVB-S standard. However, a GW can transmit a number of multiplexes. Likewise, a DVB-S system can share the bandwidth of a transponder between several GWs. The access method in this case is generally proprietary and is based on a multiplexing using time division multiplexing (TDM).

3.2.3. *IP encapsulation method over DVB-S*

3.2.3.1. *Encapsulation architecture*

The MPEG2-TS specification offers a number of alternatives in order to encapsulate private data (defined neither by the DVB group nor by the MPEG2 group) other than native MPEG2, especially IP packets. Figure 3.6 shows the possible encapsulation methods [ETS 03].

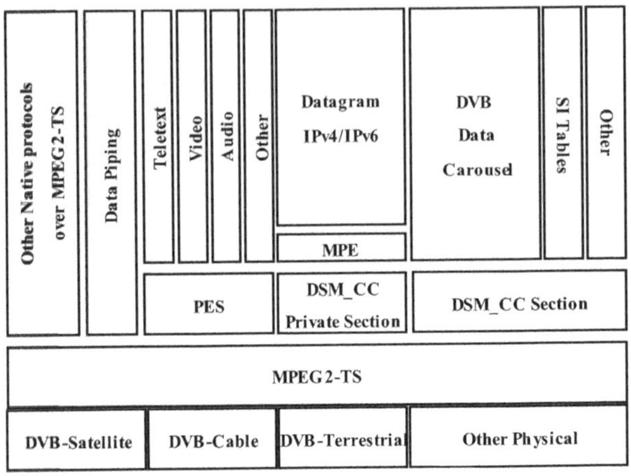

Figure 3.6. *DVB protocol stack*

The *data streaming* mode discussed previously is not particularly suitable for transporting IP packets which generally do not form continuous streams. IP packet encapsulation in PESs is nevertheless specified, in an asynchronous mode, via a PES packet data byte field, which is broken down into two types of private data (private stream 1 and 2).

The *data piping* mode gives a framework for proprietary encapsulation methods. Therefore, there is no standardized method of fragmentation/reassembly and encapsulation/de-encapsulation. The new encapsulation methods are generally based on this mode to provide alternatives, notably for the IP encapsulation in the MPE.

The *data carousel* mode consists of a cyclical emission of data toward receivers. This data is periodically encapsulated in private DSM-CC sections of fixed size called "datagram sections".

The *multiple protocol encapsulation* (MPE) mode is the encapsulation mechanism recommended by the DVB-S standard for transporting IP datagrams.

3.2.3.2. *MPE encapsulation*

MPE encapsulation provides a mechanism for the encapsulation of data above MPEG2-TS in DVB networks.

It is designed, among other things, to transport IPv4 and IPv6 packets. The MPE header contains a field which defines the address of the receiver (MAC address) of data when a number of ST addressees share the same logical channel within an MPEG2-TS multiplex. The IP datagrams are encapsulated in the DSM-CC section compatible with the format of the private MPEG2 section.

If we consider the *section-packing* mode, sections are fragmented and then inserted in order to supplement the MPEG2-TS packets. No padding bit is necessary. In order to signal the presence of a new section in an MPEG2-TS packet, a *Pointer Field* (1 byte) is placed directly behind the MPEG2-TS header. It helps to specify the number of bytes between the *Pointer Field* and the start of the new useful section.

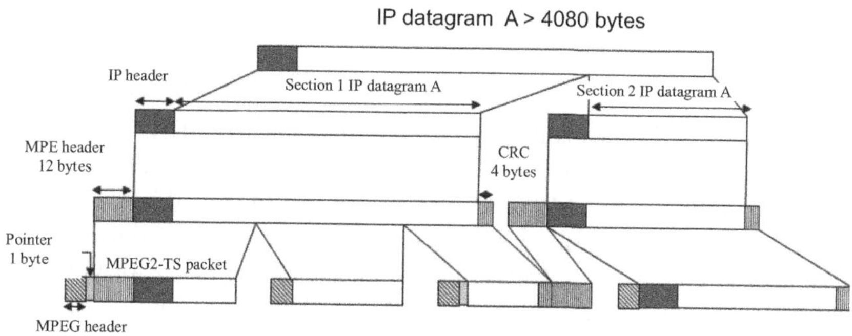

Figure 3.7. *Encapsulation of an IP datagram using MPE*

3.2.3.3. *Signaling*

The standard [ETS 04] recommends the use of the INT table (IP/MAC notification table) in order to signal the availability and location of the IP streams within a TS multiplex. The sequences of IP packets with the same source and/or destination IP address are grouped under the notion of IP stream. Using the information stipulated in the NIT, PAT, PMT and INT tables, the receiver will be able to evaluate the PID which corresponds with an IP stream or a set of IP streams. With this table, a number of descriptions enable the management of IP addresses in a DVB network notably using the following identifiers: Platform_id[4], network_id, original_network_id and transport_stream_id.

This standard remains relatively open however since it simply defines an IP/PID association table but offers no automatic processing of it by the receiver. The problem of resolving the address thus remains intact. In the case of satellites, this address resolution is a process consisting of three levels: the IP address must be resolved in a MAC address, then associated with a PID and finally with a specific TS multiplex. Whatever the level considered, the INT table does not resolve the problem of associating the PIDs according to the IP addresses. There is no association specified between the MAC and IP addresses. Finally, the INT table can only be used as part of the MPE encapsulation of the IP packets. As yet, this method is the only one to be implemented in the STs, and the association between the PIDs and IP addresses is generally configured statically.

3.2.3.4. *The ULE alternative (unidirectional lightweight encapsulation)*

Although MPE encapsulation is recommended by ETSI, it comes with limits to performance. MPE was not designed solely for IP transport. Also, a number of fields in the MPE header are not useful. For this reason, the IP over DVB group at Internet Engineering Task Force (IETF) reformed above all in order to offer an alternative to this encapsulation: ULE [FAI 05]. ULE enables encapsulation directly over MPEG2-TS of any type of PDU via a subnetwork data unit (SNDU) according to the data piping method defined by the DVB group (see Figure 3.8).

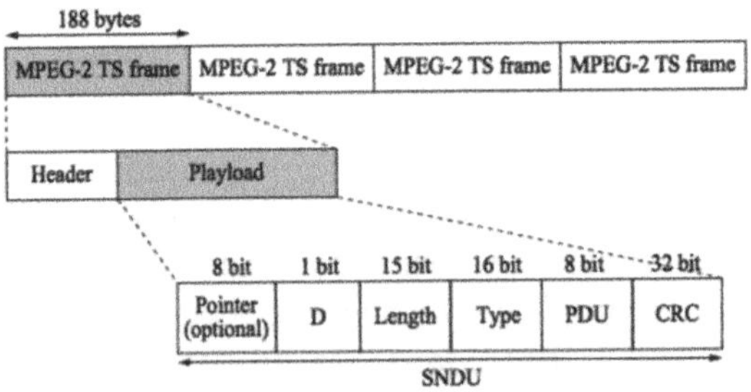

Figure 3.8. *ULE encapsulation*

4 The notion of IP/MAC platform groups IP/MAC streams and/or receivers and thus constitutes an address space without conflicts. This platform may be present in a number of multiplexes and over several DVB networks. Moreover, several platforms co-exist within a TS multiplex.

The SNDU header is made up of various fields, including:

– *indicator*: indicates the presence of a MAC destination address. This field is only optional as the receiver's filtering of packets can also be done on the PID and the IP destination address;

– *size*: defines the length of the packet;

– *type* of PDU: IPv4 (0x0800), IPv6 (0x86DD) and Bridged Ethernet (0x6558), etc.;

– *destination address* (optional): it will be inserted between the type field and the PDU. It indicates the MAC address of the SNDU receiver over 48 bits.

In order to notify the presence of a new SNDU in a MPEG2-TS packet, ULE does not require the MPEG2-TS Adaptation Field (Adaptation Field Control at 01). If a new SNDU is contained in an MPEG2-TS packet, the PUSI bit is set to 1, and ULE defines a *pointer field* inserted behind the MPEG2-TS header indicating the starting bit of the new SNDU.

Performance studies have already been carried out showing the superiority of ULE over MPE for IP encapsulation. This advantage lies mainly in the reduction of superfluous fields in the header, but it also offers a number of benefits compared to MPE regarding IPv6, the compression of the IP header and the IP multicast.

Furthermore, the IP over DVB group also proposes mechanisms for linking IP addresses to the TS multiplex [GOR 07]. Dynamic solutions are based either on the use of MPEG2-TS INTs or multicast mapping tables (MMTs), or on a real protocol for resolving the satellite IP address.

3.3. The DVB-S2 standard

Standardized by ETSI in 2005 [ETS 09g], DVB-S2 introduces new elements offering significant performance improvements compared to its predecessor at a number of different levels, mainly by altered coding and modulation, but also by an optimized encapsulation.

3.3.1. *Coding and modulations*

A major problem for satellite broadcasting systems lies in the quality of the channel which is highly sensitive to weather conditions. Variations in the propagation channel often mean that the broadcasting system needs to be dimensioned in the worst case scenario, or at least in an unfavorable scenario. As a result, this leads to an under-use of the connection due to an unsuitable coding when conditions are good. The introduction of a variable and/or adaptive broadcasting system leads to a better performance, greater flexibility and a reasonable complexity of the receiver.

In the case of variable coding and modulation (VCM), it is possible to choose the broadcasting mode from among the set of modulations of DVB-S2 (QPSK, 8PSK, 16APSK and 32APSK) as well as the type of coding (block coding and LDPC coding). In this way, a modulation and coding (ModCod) is formed from among the 28 which are available (Figure 3.9), which must be adapted to the quality of the connection (C/N). Thus, the ModCod QPSK ¼ enables broadcasting over a degraded connection (C/N = -2.35 dB) with low efficiency (0.36 spectral efficiency) and, conversely, the ModCod 32APSK 9/10 will be the most efficient (3.3 spectral efficiency) over a very high quality connection (C/N = 16.05 dB).

Mode	MODCOD	Mode	MODCOD	Mode	MODCOD	Mode	MODCOD
QPSK 1/4	1_D	QPSK 5/6	9_D	8PSK 9/10	17_D	32APSK 4/5	25_D
QPSK 1/3	2_D	QPSK 8/9	10_D	16APSK 2/3	18_D	32APSK 5/6	26_D
QPSK 2/5	3_D	QPSK 9/10	11_D	16APSK 3/4	19_D	32APSK 8/9	27_D
QPSK 1/2	4_D	8PSK 3/5	12_D	16APSK 4/5	20_D	32APSK 9/10	28_D
QPSK 3/5	5_D	8PSK 2/3	13_D	16APSK 5/6	21_D	Reserved	29_D
QPSK 2/3	6_D	8PSK 3/4	14_D	16APSK 8/9	22_D	Reserved	30_D
QPSK 3/4	7_D	8PSK 5/6	15_D	16APSK 9/10	23_D	Reserved	31_D
QPSK 4/5	8_D	8PSK 8/9	16_D	32APSK 3/4	24_D	DUMMY PLFRAME	0_D

Figure 3.9. *Set of ModCods available in DVB-S2 (source ETSI)*

The standard advocates a dynamic adaptation of ModCods according to the C/N measured by the satellite terminal and the dialogue with the gateway. This type of broadcasting is called adaptive coding and modulation (ACM), since it allows a continuous adaptation of the ModCod in order to offer the best efficiency. Given the different coding and modulations used for each connection, the bandwidth attributed to each terminal may therefore vary depending on the error rate and thus on the modulation and coding given.

Figure 3.10 shows the adaptation of the ModCod depending on the quality of the signal measured by the terminal. These measures were carried out on a DVB-S2 terminal and a gateway coupled by a channel emulator. The measure consisted of periodically varying the attenuation (visible on the second curve) over a time scale (visible on the first curve) of 2 min. The second curve indicates the value of the energy-per-symbol of the noise density (Es/No) measured as well as the ModCod chosen by the system. In these conditions, the ModCods evolved from QPSK ¼ to 8BPSK ¾.

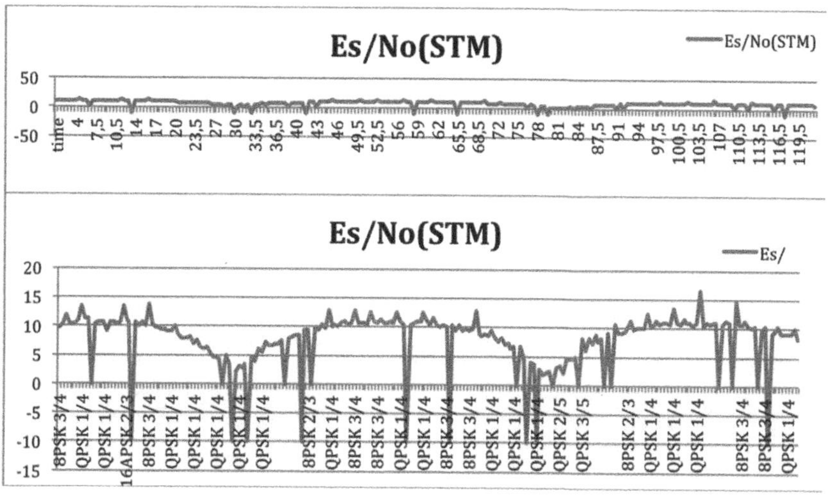

Figure 3.10. *SNR and ModCod vs. time to noise*

3.3.2. *Encapsulation*

Among the improvements provided by the DVB-S2 standard, the generic stream encapsulation (GSE) protocol [ETS 07, ETS 11b] has a significant impact on the architecture of the system. In order to optimize the encapsulation of the data traditionally carried out by MPEG2 transport stream/multiple protocol encapsulation (MPEG2-TS/MPE) which is seen as being inefficient, DVB-S2 offers a new encapsulation diagram based on the GSE protocol aiming to minimize the additional cost of encapsulation.

GSE data connection level protocol for transporting packets comes from a number of protocols (IPv4, IPv6, MPEG, ATM, Ethernet, etc.) directly above the physical layer (DVB-S2, DVB-T2, DVB-C2). High-quality fragmentation mechanisms enable the transmission of IP packets (or others) above a physical layer possibly using adaptive or variable coding and modulation mechanisms (ACM or VCM). A 32-bit CRC is added at the end of a sequence of fragmented packets solely in order to ensure the integrity of the reassembly and not the integrity of the data.

Figure 3.11 shows the DVB-S2 principle of IP encapsulation with GSE (source ETSI)

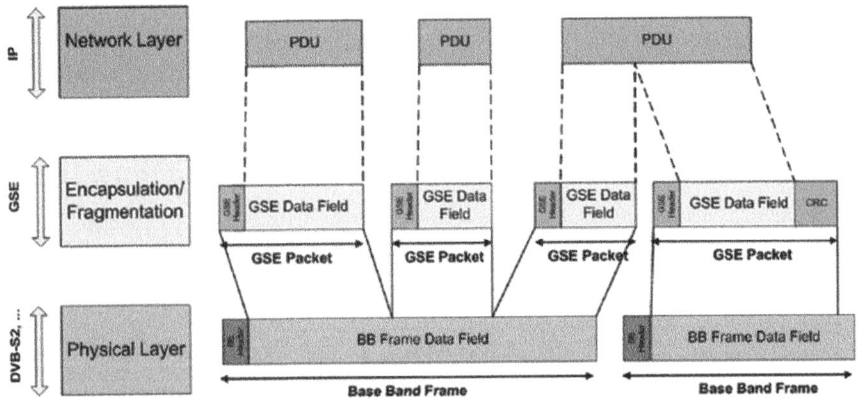

Figure 3.11. *Diagram of IP encapsulation over DVB-S2 by GSE (source ETSI)*

The objective of GSE was to achieve an additional encapsulation cost of below 3% for a normal-sized packet. Compared to MPE, GSE reduces the transmission cost by a factor of 3 and increases the capacity by 10%.

3.4. The DVB-RCS standard

With the DVB-S standard, the terminals are only able to receive traffic coming from the satellite. As the IP services (broadband Internet access, voice over IP, on-demand video, etc.) have grown to the scale of terrestrial networks, the idea of supporting an interactive channel between the user terminal and the servers and Internet users emerged. The intrinsically asymmetrical nature of this dialogue and the requirement of a competitive access to the return channel in order to prevent a monopolization of the bandwidth by a single station, excludes the use of the DVB-S standard. This standard does not define any media shared access methods and requires equipment which is too expensive for the end user.

Thus, the unidirectional link routing (UDLR) standard [DUR 01] helped to provide a bi-directional solution at a lower cost via a terrestrial return link, generally a traditional RTC modem link.

However, the satellite network linked to a terrestrial infrastructure ceases to be fully autonomous, hence the need for a satellite return channel. The DVB-RCS standard as specified in [ETS 09a, ETS 09c] standardizes this satellite return channel for satellite terminals. Combined with the forward channel using the DVB-S standard, DVB-RCS introduces the interactivity necessary for the definition of the architecture of a next-generation broadband multimedia system.

The DVB-RCS standard is the cornerstone of bi-directional satellite systems since it enables the STs to interact. As we have seen previously, this capacity to interact requires the support of a new resource access model over an uplink.

3.4.1. *Access method: MF-TDMA*

The resource access model in the return channel is based on multiple frequency-time division multiple access (MF-TDMA). This system relies on a frequency decomposition of the return link into several frequency bands, which are shared into time slots, during which the STs are transmitted in bursts of data.

3.4.1.1. *Different RCS bursts*

There are four types of DVB-RCS burst:

– bursts of traffic based either of ATM cells (ATM TRF (traffic) burst), or MPEG2-TS packets (MPEG2-TS TRF burst);

– the common signaling channel (CSC) burst allowing the ST to identify itself with the NCC during the logon phase;

– the ACQ (acquisition) burst used, if necessary, during the synchronization procedure;

– the SYNC (synchronization) burst used to maintain the synchronization or send control information to the NCC.

A burst of ATM cells is thus made up of a combination of ATM cells (the number of which depends on the size of the time-related transmission unit), an optional field called satellite access control (SAC) and an obligatory header.

This SAC field can also be found in the SYNC burst, periodically sent during the ST connection in order to keep it synchronized.

The other ACQ and CSC bursts have their own encapsulation syntax.

3.4.1.2. *Segmentation of resources: time-slots, frames and superframes*

The distribution of timeslots of a MF-TDMA channel is centralized by the NCC which periodically allocates to STs a series of bursts. Each one is defined by a frequency, a bandwidth, a start time and a duration.

Time-slots are organized into frames, which are then grouped into superframes. A superframe delineates the set of resources of a DVB-RCS transponder shared by a group of STs. Figure 3.12 shows the composition of a DVB-RCS superframe.

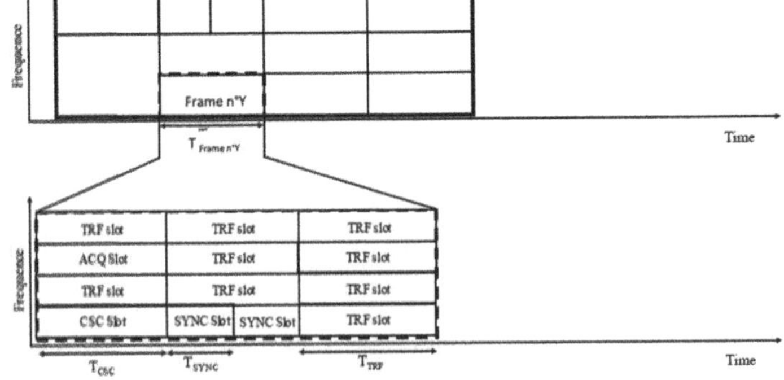

Figure 3.12. *Composition of a DVB-RCS Superframe*

Each superframe is associated with a plan to allocate time-slots reserved for each ST in the rest of the frame. The frames within a superframe do not necessarily have the same duration or division into time-slots.

Two modes of MF-TDMA sharing are defined: one mode is based on fixed time-slots (static MF-TDMA) and the other mode is based on dynamic time-slots (dynamic MF-TDMA). In the static mode, the duration and the bandwidth of the successive time-slots used by an ST are fixed. A modification of their characteristics can only be applied in a new superframe. In the optional dynamic mode, the bandwidth and duration of time-slots allocated to an ST may vary within a superframe. Thus, the ST, as well as adapting to the change in carrier frequency and duration of time-slots, may also need to modify its transmission rate and the type of coding it uses from one burst to another. This mode enables it to adapt as best as possible the characteristics of the time-slots to the various profiles of multimedia traffic as well as to the propagation conditions. However, this increased flexibility comes at the expense of access time to the support and it requires extra calculation capacities from the NCC and the STs.

3.4.1.3. *The protocol stack in the data plan*

The DVB-S/RCS systems are based on the forward channel on DVB-S and for the return channel on DVB-RCS. On the forward channel, three protocol architectures including IP are permitted:

– IP in AAL5/ATM (according to the specification [GRO 99]) in a burst of traffic);

– IP in AAL5/ATM in MPEG2 TS (according to the specification [ETS 99] in a burst of traffic);

– IP in MPE in MPEG2-TS in a burst of traffic (optional method defined in [ETS 09a]).

On the return channel, IP can be encapsulated in bursts, either standard ATM or MPEG2-TS. The protocol architecture shown in Figure 3.13 is the architecture adopted if a regenerative satellite is used.

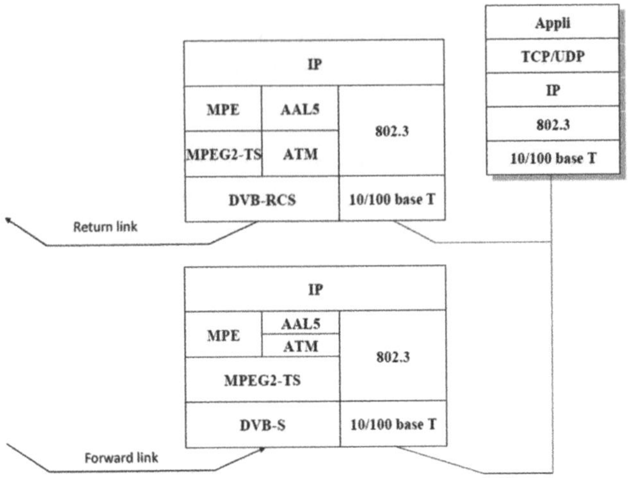

Figure 3.13. *DVB-S/RCS Protocol Architecture in the Data Plan*

3.4.2. *Signaling in a DVB-RCS/S System*

3.4.2.1. *Signaling on the forward channel: specific DVB-RCS tables*

SI tables sent in the MPEG2-TS transport stream on the forward channel contain the set of parameters to allow the receiver to demultiplex and decode the various streams in the multiplex. This data is structured in tables encapsulated in private DSM-CC sections, using MPE, and transported in logical channels to predefined PIDs. The table sent to STs can be divided into two categories:

– the PSI/SI tables detailed in section 3.2.1.2;

– the new SI tables necessary for the correct operation of the DVB-RCS protocol: superframe composition table (SCT), time slot composition table (TCT), satellite position table (SPT), frame composition table (FCT), terminal burst time plan (TBTP), correction message table (CMT).

Therefore, the SCT describes the complete partitioning of satellite resources into superframes and the relative positioning of the frames in relation to one another within superframes. For each frame, the FCT table contains its decomposition into time-slots and its respective location. Each of these time-slots is described by the TCT which contains all the technical characteristics of the time-slot and the type of burst contained (ATM or MPEG2-TS TRF, SYNC, ACQ and CSC). Finally, the TBTP table contains the plan for the allocation of time-slots in a superframe shared between a group of STs. The SPT table indicates the position of the satellite and the CMT table enables the STs to carry out the necessary corrections for the next transmissions.

Terminal information messages (TIMs) are private messages sent by the NCC to the ST or to a group of STs. They are used to transport logical identifications (Group_id, Logon_id, Channels_ids, ST IP addresses, PID values, PVCs), and success or fail messages for the logon procedure.

Finally, the network clock reference (NCR) allows the STs to synchronize themselves.

The set of frames is listed in Figure 3.14.

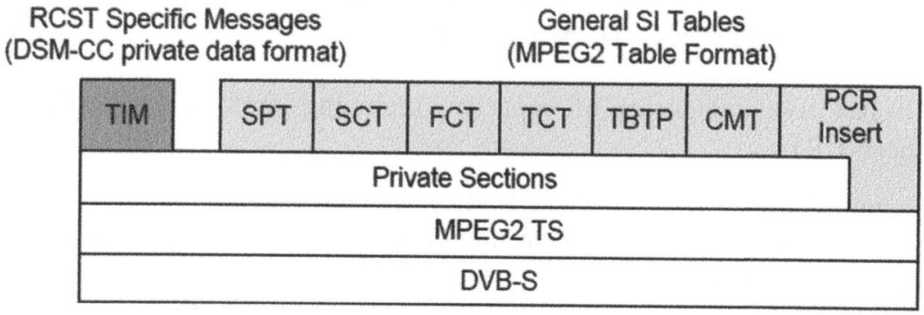

Figure 3.14. *Protocol stack for RCS signaling on the forward channel*

3.4.2.2. *Signaling on the return channel*

3.4.2.2.1. Logon and synchronization procedures

At initiation, the ST starts a logon procedure in order to access resources of the satellite link. The terminal satellite transmits an initial access request by CSC burst to the NCC. This burst includes the MAC address of the ST and information regarding its capacities. After the acceptance from the NCC, it responds with a TIM message containing the logical identifications associated with the ST: virtual path identifier (VPI) and virtual connection identifier (VCI), ATM and PIDs to transmit traffic and exchange control and administration messages (MPEG2-TS burst). The NCC also allocates it a time-slot dedicated to the synchronization signaling, which will notably allow the terminal to keep the synchronization of the NCR reference clock, via periodic SYNC messages.

The information exchanged during the logon phase can be used in order to fill in the database in the NCC and the ST, simplifying the address resolution. The allocation of an IP address and the IP address of the GW is by default not specified by the standard and thus generally depends on manual configuration or SNMP.

3.4.2.2.2. Categories of capacity requests

One of the main advantages of the DVB-RCS standard is that it offers a return channel whose capacities can vary over time. Each ST can dynamically request "transmission capacities" to the NCC which allocates them depending on the requirements and resources. Therefore, if at any time there is an accumulation of traffic observed in an ST, it may require more bandwidth from the NCC. The corresponding algorithm is the demand assigned multiple access (DAMA) (see below).

It has five categories of capacity, including three types of capacity request (CR):

– continuous rate assignment (CRA): a fixed quantity of time-slots per superframe negotiated at logon to the ST is allocated for the duration of the connection to the ST;

– rate-based dynamic capacity (RBDC): a quantity of time-slots is negotiated for a superframe but cannot exceed a maximum threshold ($RBDC_{max}$). Although it is optional, this capacity can punctually supplement a minimum capacity allocated by the CRA;

– volume-based dynamic capacity (VBDC): a quantity of time-slots negotiated with the NCC may be distributed between several superframes. These requests are cumulative;

– absolute volume-based dynamic capacity (AVBDC): a quantity of time-slots negotiated with the NCC may be distributed between several superframes. A new AVDBC requests cancels out the previous request;

– free capacity assignment (FCA): this type of capacity represents the time-slots of a superframe which remain after the processing of an other type of capacity by the NCC. They are distributed, either evenly or unevenly, between the concurrent STs to the level of a predefined threshold.

Time-slots are distributed to each superframe and this allocation complies with the following priority scale between the types of capacity: CRA>RBDC>A(VBDC)>FCA.

3.4.2.2.3. DAMA

Demand assigned multiple access (DAMA) is a client/server type protocol enabling, in a DVB-PCS system, an allocation based on the demand for resources at MAC level. The client and server are located in each ST and in the NCC, respectively. Signaling of STs toward NCC may be, for example, out-of-band or it may be carried out via the SAC prefix in SYNC bursts. The NCC responds by sending the TBTP table to each superframe.

DAMA can be broken down into six distinct phases:

– the calculation of CRs in the terminals;

– the transmission of CRs toward the MAC scheduler;

– the calculation of the MAC scheduler of the resources to be allocated;

– the distribution of these resources within the superframe;

– the generation and transmission of TBTP defining the allocation plan of the time-slots;

– distribution inside STs of these resources between end users and their applications.

Currently, the particularly open character of the DVB-RCS standard on DAMA promotes proprietary solutions which are difficult to access.

In order to transport CRs toward the NCC, two types of signaling are supported: intra-band and out-of-band signaling.

Out-of-band signaling is based on a method linked to the use of the optional SAC field attached to the traffic burst, or on the data unit labeling method (DULM) method, which enables the ST to send control and/or administration information to the NCC in bursts normally dedicated to traffic.

Intra-band signaling is based on the "mini-slot" method with or without conflict, which corresponds to the periodic allocation to an ST or a group of STs with a burst of a shorter duration than traffic bursts.

This allocation of bandwidth depending on demand is unquestionably one of the strengths of satellite systems in terms of QoS, obtained by distributing the total satellite bandwidth between the STs according to their potential load.

3.4.3. *Connections*

In appendix J of [ETS 09c], the notion of Connection Control Protocol (C2P) is developed. As well as the various types of connections possible (uni- and bi-directional, point- or multipoint-to-point or multipoint-to-multipoint), the different identifiers are detailed (Channel_id, Source & Destination Address, Forward and Return Stream Identifier).

In the AmerHis project, three types of connections are distinguished in order to respond to requirements of the different types of services and to various modes of resource management:

– permanent connections, established at logon for the duration of the connection to the ST with basic characteristics (CRA, $RBDC_{max}$, etc.), defined by the intermediary of SLAs which may be renegotiated during the connection to STs;

– semi-permanent connections, negotiated at logon to the ST, but whose activation is deferred in time to satisfy the requirements of periodic services;

– connections on demand, involving a dynamic request from the ST, requiring admission control and resource reservation with the NCC. These connections are used for one-off services.

Therefore, C2P may be used to modify the parameters of a permanent connection or to require an on-demand connection. In no way does the DVB-RCS standard solely specify C2P. Currently, ST requests to the NCC are coded in the information elements (IE) fields transported in the DULM format, and they establish/modify/close a connection or channel. The NCC responds by various descriptors of TIM messages. However, alternatives to this level 2 signaling are considered via protocols at higher levels.

3.5. DVB-RCS2

The DVB-RCS2 standard was published by ETSI in March 2011 [ETS 11a]. Almost 10 years after the DVB-RCS standard, DVB-RCS2 supplemented the progress made in the forward channel (DVB-S2) by optimizing the transmission and encapsulation techniques.

The standard is divided into three distinct parties: RCS-lower layer satellite (LLS) which defines a specification for an interactive DVB-RCS2 return channel, which may be seen as an improvement to the previous generation at the level of the physical layer, the media access control (MAC layer) and IP encapsulation. RCS-high layer for satellite (HLS) defines a support for protocols and application based on IP, offering interoperability and better management of resources. Finally, RCS-OSL offers combinations of technologies and commercial profiles in order to respond to the needs of the market (small/home office, corporate, military, backhauling, etc.)

3.5.1. *Coding and modulation*

One of the most highly anticipated extensions is evidently the use of adaptive coding and modulation (ACM). DVB-RCS2 integrates optional ModCods 8PSK and 16AQM, as a supplement to the traditional QPSK. It also uses a new forwarding error correction (FEC) algorithm with a turbo code of 16 states for TDMA bursts, offering up to 2 dB of extra gain. The combination of forward and return channels delivers a real gain in terms of transmission, but also represents a significant challenge to operators in terms of its configuration. Even if the adaptation may be technically guided by the application, it is likely that it will be managed by the system with a view to maximize the transfer rate while retaining acceptable levels of QoS and drop rates.

Alternative (nonlinear) modulations are also proposed (CPM – continuous phase modulation). They enable the production of cheaper terminals which are also less efficient.

3.5.2. *Access techniques*

A new random access technique supplements DAMA in the return channel in order to prevent a prior request for capacity via the traditional allocation cycle. Thus, in the case of small transfers, the system is much more reactive, but demands which are too high can also lead to a reduction in the efficiency of the allocation mechanism before reaching saturation point. This random access (RA) technique may be seen from the terminal end as a semi-static low transfer-rate allocation. The number of time-slots which a terminal can access in the superframe using this method remains relatively constant over a short period of time (tenths of a second). The terminal randomly accesses a selected pool of slots, which may evidently lead to collisions and therefore drops. This technique is therefore a compromise between the lack of reactivity of an efficient access mechanism (DAMA) and the low efficiency of constant rate assignment (CRA).

3.5.3. *Encapsulation*

In the same way that DVB-S2 optimized encapsulation methods, DVB-RCS2 is also optimized for directly transporting the IP on the satellite. A standard GSE protocol can also be used on the forward and return channels, since return link encapsulation (RLE) is an adaptation of GSE for the return channel.

3.5.4. *QoS architecture and PEP*

The QoS architecture defined in DVB-RCS was also modified. This architecture resembles the one defined by the BSM group at ETSI. Cross layer interactions between the lower layers (allocation) and higher layers (IP DiffServ) enable a better use of the system. The correspondence of service classes with physical resources is one of the key points of the architecture. This architecture also takes account of the new access mode in the channel (RA).

The high layer (HLS) introduces a performance enhancing proxies (PEP) negotiation protocol for improving TCP performance on the satellite. Therefore, according to the standard, a satellite terminal can select an available PEP in the network after a discovery phase.

3.6. QoS architecture in DVB-S/RCS satellite access networks

When resources are limited, the bandwidth of satellite systems must be optimized. To do this, in the DVB-S/RCS standards, ETSI defines solutions on a physical level for spectral optimization, but does not define the QoS management architecture (which is nevertheless necessary for an optimal use).

Two supplementary approaches offer architectural frameworks. The first approach, upheld by SatLabs, can be considered as a short-term approach since it focuses on the return channel of the satellite system. The second approach, from ETSI BSM, adopts a more prospective approach. Before addressing these two architectures, we will present the stakeholders competing to supply Quality of Service (QoS) in a satellite network proposed by the BSM [ETS 01].

3.6.1. *The various stakeholders in the satellite network*

One of the key aspects of integration of a network into NGNs is to properly separate the various stakeholders competing to supply a service in the network. For satellite networks belonging to traditional telecommunication networks, this process becomes all the more significant given that these networks are still highly dependent on their close vertical integration.

These stakeholders were clearly identified in the technical report of BSM ETSI TR 101 984 [ETS 01] on the reference services and architectures for BS systems:

– the *satellite network operator (SNO)*: possesses and manages the maintenance, management and deployment of the DVB-RCS system, apart from the terminals (STs & GWs). It also manages the division of resources between INAPs depending on their contract;

– the *satellite operator (SO)*: is responsible for the satellite and cooperates, for a regenerative system, with the SNO for the configuration of the OBP;

– *interactive network access provider (INAP)*: has access to some of the satellite resources via an interactive network (IN) which it negotiates with the SNO. It then shares these resources between various SPs via virtual satellite networks (VSN). INAP are also often called virtual satellite network operators (VSNO). A VSN represents the resources offered by an INAP to an SP. The choice however lies with the SP of whether or not to manage these satellite resources. We can therefore distinguish two types of contracts named, respectively, "bandwidth" where the resources attributed are entrusted to the ST, and "wholesale connectivity" where the management of resources is the responsibility of the INAP and distributes these resources among several SPs;

– *service providers (SP),* which may be ISPs, but also application service providers (ASPs), multicast service providers (MSP), VPN service providers (VPN SP), internet telephone service providers (ITSP), etc;

– *subscribers*: are the intermediate entity to whom users can delegate the choice of SPs. A subscriber can use the services made available by INAP and agree various subscriptions with different SPs in order to satisfy users' demands;

– *users*: may be directly linked to the satellite network by the ST or linked by a LAN connected to this ST. User terminals are connected to applications provided by different SPs. The group of the LAN and the user terminals is referenced under the name of customer premises equipment (CPE).

3.6.2. *The SatLabs architectural model*

SatLabs[5] is an association grouping the stakeholders of the space sector involved in the DVB-RCS standard. It aims to promote the RCS standard, mainly via interoperability work and certification. In a general sense, recommendations made by SatLabs can be considered as "short-term", but are nevertheless coherently defined alongside the "long-term" visions of the BSM.

The main SatLabs recommendation in the QoS domain is [SAT 10a] which defines the QoS architecture and the associated specification. It should be noted that only the return channel (DVB-RCS) is addressed in this document.

3.6.2.1. *QoS architecture*

Reference architecture for QoS is defined in Figure 3.15.

5 SatLabs: http://SatLabs.org. The cited documents can be consulted for free on the website.

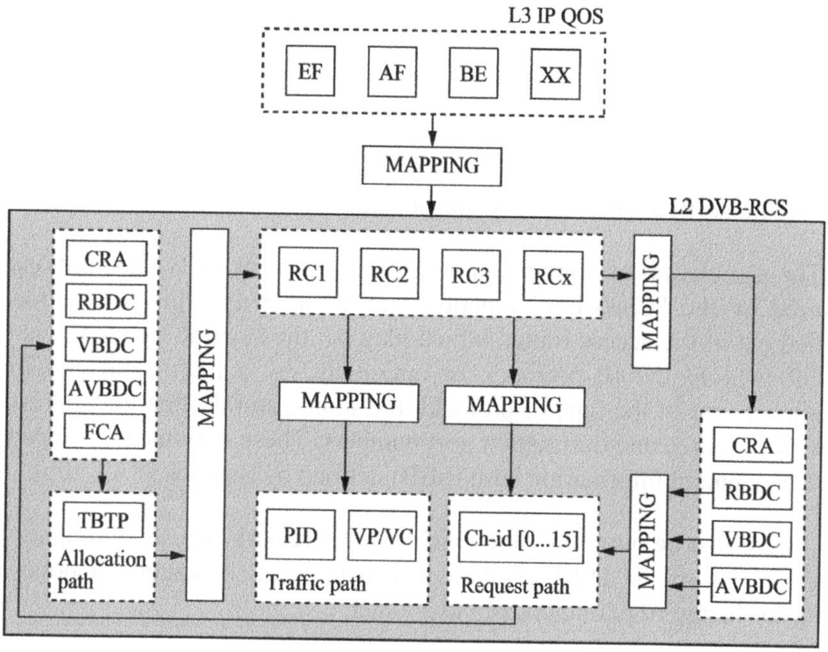

Figure 3.15. *QoS architecture DVB-RCS SatLabs (source SatLabs)*

It is clear that this architecture mainly takes account of the QoS at the network level (IP), the QoS at the access level (MAC) and the associations and interaction between these two layers.

3.6.2.2. *QoS in the IP layer*

QoS management at the IP level conforms to DiffServ, with terminals which can be either ingress nodes, or core nodes in the DiffServ domain. A DVB-RCS network can be considered as a DiffServ domain (i.e. a set of nodes ensuring a common QoS policy). In this context, the ST is one of the ingress node routers.

The different PHBs which need to be supported by a SatLabs-compatible system are:

– expedited forwarding (EF), optionally;

– assured forwarding (AF), optionally:

- at least one AF class must be supported,

- at least two "drop precedences[6]" must be considered,

- PHBs AF31, AF32 and AF33 are recommended;

– best effort (BE).

Additional PHBs may also be considered.

The association of application streams at different PHBs is traditionally ensured by the "Classifier" function (see Chapter 1). This association is carried out in the access router, which may be: the satellite terminal, via the DSCP field of the IP protocol, or optionally, by a multi-field classifier, taking account of the type of protocol, the source and destination addresses, as well as source and destination port numbers. These options are mapped in the management information base (MIB) defined by SatLabs [SAT 10b].

It may also be noted that the DSCP can be remarked if the terminals are egress nodes of the DiffServ domain and the QoS management policy is likely to change from one domain to another.

The performance information shown in Table 3.1 can be considered as a reasonable example of the system deployed.

PHB	Delay		Jitter		Priority	Bandwidth		Packet loss ratio	
	Nominal	Over-load	Nominal	Over-load		Nominal	Over-load	Nominal	Over-load
EF	Sufficiently low (no overbooking)		Minimum possible			Given in full during session, when admitted			
	300 ms		50-100 ms					< 0.1%	
AF31 /AF32	As good as possible				High	In contract			
	850 ms	larger for out of profile traffic						< 0.1%	
BE	Not controlled		Not controlled		Low				
	850 ms	larger for out of profile traffic							

Table 3.1. *Reasonable performance of a deployed DVB-RCS network*

6 Priority levels in DiffServ terminology.

It should be noted that this performance information relates to the return channel, and that many factors can influence performance of this segment, whether on the ST, NCC or capacity management level.

Above all, performance in terms of delay and jitter are particularly affected by ST performance, the traffic load and the slot allocation policy at the NCC (see section 3.6.2.3). It should be noted that the latter is particularly dependent on the overall demand of resources, the allocation policy applied to the ST and reservation mechanisms implemented in the ST.

With a view to the targeted performance, it should be recalled that the use of a satellite network has a significant impact on end-to-end performance. The characteristics of one of its PHBs will certainly dominate the end-to-end service.

3.6.2.3. *QoS in the MAC layer*

In a DVB-RCS system, the QoS functions are taken on by queues and the associated scheduler on the one hand and by resource management on the other hand.

SatLabs defines the notion of request class (RC) which is the equivalent of PHBs at the access level. An RC groups both the queues and the means for managing the allocation on the return channel. A correspondence of PHBs at the IP level must be carried out with the RCs. In general, several PHBs may be associated with a single RC.

Three RCs have been defined:

– real-time, optional, with which the PHB EF is associated;

– critical data, optional, with which the PHB AF is associated;

– best effort, with which the PHB BE is associated.

At the resource management level, each RC can be associated with requests of RBDC (data-transfer rate), AVBDC or VBDC (size), or both. Each RC must be identified by a Channel-Id DVB-RCS. The corresponding CRs contain the information on Channel-Id, allowing the NCC to recover the RC in order to carry out the allocation (following the policy implemented by the manufacturer or operator).

Continuous allocations (standard CRA) or residual allocations (standard FCA) supplement the resource management device.

For instance, an ST can implement two RCs which use the VDBC allocation, and manage the priority of one of the classes over the other.

It should be noted that SatLabs specifies the various possible allocation categories as well as the associated parameters.

Finally, regarding the addresses themselves, whether they correspond to type A or B terminals (ATM or MPEG), SatLabs only specifies the possibility of associating one or more RCs with a VPI/VCI or PID. However, in the case where PHBs EF is managed, 2 VPI/VCI (or PID) must be set out, one of which is associated with the RT traffic.

3.6.2.4. *VoIP in satellite systems*

SatLabs also addresses aspects related to the QoS, notable via work on VoIP support in satellite telecommunication systems [SAT 10, EMS 04, NER 04, IPT 05]. The majority of the documents examine the means to be implemented at the access level in order to obtain sufficient performance for transporting the VoIP. These notably include the various DVB-RCS allocation methods (CRA, RBDC and VBDC) and associated performance. Nevertheless, none of the recommendations reach a conclusion on the type of allocation or configuration to implement. [IPT 05] goes further by proposing a bibliography of the stacks, used codecs, as well as the signaling protocols (SIP, H323 and SKINNY) and recommendations. Thus, the study concludes that the CRA and the RBDC are relative equivalents when the RBDC allocation is carried out without congestion, even if background traffic can significantly affect delays in all cases. SIP and H323 are advocated, and the authors focus on the issues of network architecture (e.g. DNS), and the impacts on general performance of transfer rates and sampling frequencies of codecs.

Generally speaking, this work clearly illustrates the issues involved in interactive multimedia applications in a satellite context for a given application. Obviously, a generic approach which can guarantee the requirements of QoS in a heterogeneous environment can be based on the information provided in these studies.

3.6.2.5. *Example of QoS architecture for the return channel*

This example demonstrates the architecture of a DVB-RCS satellite terminal on the market made by the manufacturer STM. The STM Satlink 1000 satellite terminal [STM 07] is one of the SatLabs certified DVB-RCS terminals, supporting DVB-S and DVB-S2 standards on the forward channel. Optimized for IP networks, it targets Internet access services for individuals and small businesses. It has functions for QoS management, traffic acceleration, VPNs, as well as multicast applications and the support of the DVB-S2 standard implementing ACM.

For QoS management on the return channel, the STM Satlink 1000 ensures a differentiation of service based on DiffServ architecture and offers per hop behavior (PHB) performance as defined in the DiffServ architecture.

The STM classifies the traffic which will be sent on the return channel in different QoS groups. One or several QoS groups are then associated with a PHB or a group of PHBs. The QoS groups supported in the STM are shown in Figure 3.16.

QoS Group ID	QoS Group Name	Maps to PHB
0	Best Effort	Best Effort (BE)
1	VoIP Audio	Real Time Voice over IP (RT-VoIP)
2	VoIP Signaling	Real Time Voice over IP (RT-VoIP)
3	ViC Video	Real Time Video Conferencing (RT-ViC)
4	ViC Audio	Real Time Video Conferencing (RT-ViC)
5	ViCSignaling	Real Time Video Conferencing (RT-ViC)
6	Critical Data	Critical Data(CD)

Figure 3.16. *The QoS groups supported by the STM SatLink 1000*

The routing of each IP packet toward the corresponding QoS is based on a multi-field classifier, such as DSCP/TOS fields or other IP fields (source and destination addresses and ports, type of protocol).

At the MAC level, four queues are defined:

– best effort (BE);

– critical data (CD);

– real time video conferencing (RT-ViC);

– real time voice over IP (RT-VoIP).

The RCs are differentiated by adapted management of:

– the algorithm of allocation requests: the *real-time* classes are fed by CRA requests. The *critical* and *best effort* classes are dimensioned to receive more variable traffic which is less sensitive to delay, based on A/VBDC;

– the transmission of PIDs or VPI/VCI is used to define a priority for the real-time traffic at the MGEP/ATM level, when concurrent traffic is intertwined;

– the queue management policy: in the event of an excessive load, and thus an overflowing of the queues, the real-time packets are dropped at the head of the queue (the oldest packets) while the best effort traffic is dropped from the back of the queue (newly arrived packets).

3.6.2.6. *Example of QoS architecture for the forward channel*

Although the forward channel is not addressed by SatLabs, it is not difficult to ascertain the architecture outline. The uniqueness of the QoS in the forward channel, which also contributes to its simplicity, is that it only relies on IP level mechanisms. As shown in Figure 3.17, taken from the AmerHis project, the packets coming from ISP1 and ISP2 firstly undergo a static admission control in the border router in order to verify that the traffic added by each ISP in the network complies with the SLA negotiated between the INAP and the ISP. The signaling traffic is marked and limited according to the filtering rules defined in the INAP-SP SLAs. The packets are then multiplexed (IP Mux) and scheduled before being sent to the GW.

At the GW level, there is a DiffServ server consisting of a classifier, a spacer and a regulator. The IP packets are again rescheduled because they are multiplexed on the forward channel with the DVB-S and DVB-RCS signaling. For this reason, a control mechanism for the output rate of the IP scheduler needs to be implemented in order to adapt the quantity of MAC signaling generated. We should recall that the GW transmission rate on the forward channel is constant. The IP packets are then inserted into the MPEG2-TS logical channels where the packets remain ordered. The capacity of the channel corresponds to the aggregation of requirements of each receiving subscriber, by service class, regardless of the ISP.

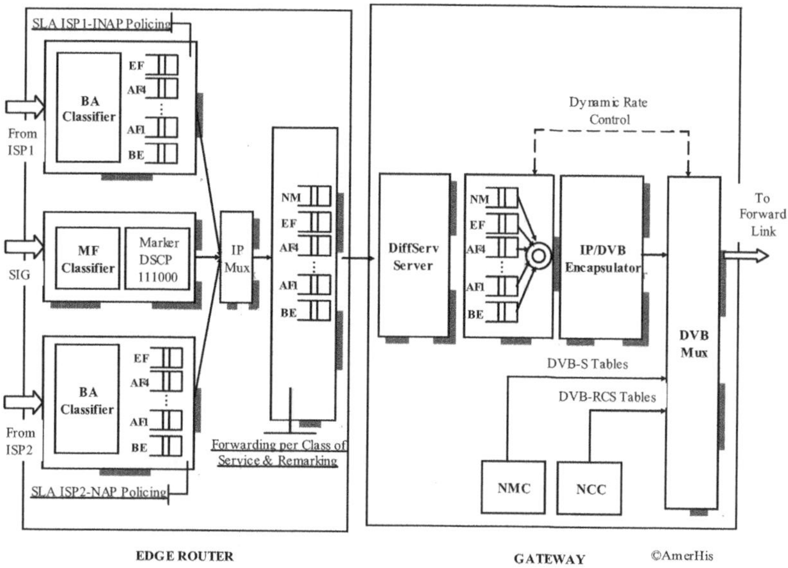

Figure 3.17. *QoS in the edge router and the gateway*

In conclusion, the SatLabs recommendations on QoS constitute a working basis for QoS mechanisms in a DVB-RCS system, by offering a common architecture as well as a more detailed definition particularly of the aspects related to allocation and addressing. They enable a better interoperability of equipment sharing a common framework.

Nevertheless, the specification and definitions are not restrictive and precise enough to unify the QoS management in this type of system.

3.6.3. *The BSM architectural model based on IP*

Via a technical specification produced by the ETSI technical committee "Satellite Earth Stations" (SES), BSM defines a QoS architecture for next-generation networks.

In general, this work differs from that produced by SatLabs in that it takes a more prospective approach regarding the implementation of mechanisms which are more evolved than DiffServ (and their demodulations at access level) and that it takes account of the architecture inherent to BSM. This (Figure 3.18) is characterized by the separation between the satellite independent layers (known as SI) and those which are dependent on the satellite architecture (SD layers – satellite dependent) and thus subject to different implementation by manufacturers. Communication between these layers is carried out by the intermediary of access points of generic service known as SI-SAP.

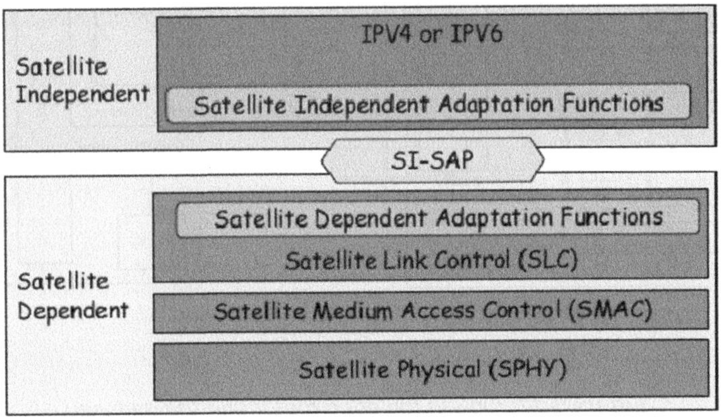

Figure 3.18. *BSM architecture*

3.6.3.1. *Context and outline*

Unlike the work of SatLabs regarding QoS, BSM not only examines the satellite segment, but also the aspects related to interconnection and end-to-end QoS[7].

Therefore, [ETS 06] introduces the traditional differentiation between end-to-end teleservices (including users' terminals) and the network support services which are limited to the network (excluding the terminals this time). These support services provide a means of communication to UNI/NNI interfaces and correspond to layers 1 to 3.

As shown in Figure 3.19, the IP support service is defined between the border edges (from UNI to UNI). It is itself based on the support services of the different segments crossed (including the satellite network) and the core

7 SatLabs briefly address this issue via associations of traffic classes at the border of the DiffServ domain.

network. Obviously, the QoS provided from end-to-end is based on the IP support QoS, which itself is restricted by the QoS of various subsegments, as well as by the QoS provided by the network user (CPN in the figure below).

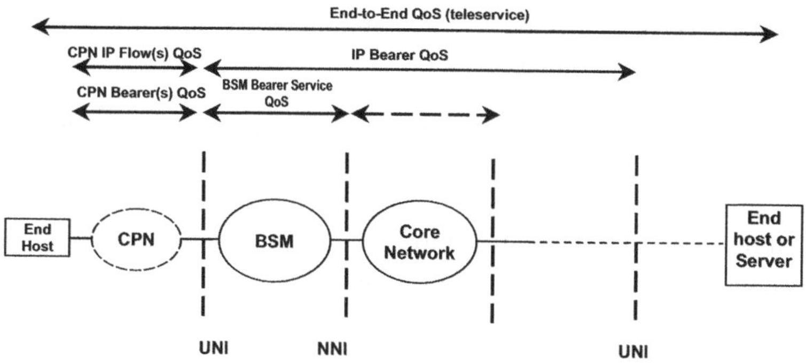

Figure 3.19. *Overview of the BSM QoS architecture*

3.6.3.2. *QoS models*

Two QoS models are put forward in the BSM QoS functional architecture:

1) a model of guaranteed QoS, which enables compliance with the values of QoS parameters (such as transfer rate, maximum delay, jitter, drops, etc.). This model can be associated with the IETF's IntServ model;

2) a model of relative QoS, without an absolute guarantee of compliance with QoS criteria, but which allows differentiated processing between service classes. This model can be associated with the IETF's DiffServ model.

Nevertheless, a combination of these two models is advised, both from a point of view of the QoS offered to certain interactive application services such as VoIP, and from a point of view of the heterogeneity of the segments crossed or the issues linked to scaling up. In this case, a guaranteed QoS model could be used for a limited number of traffic types and only in restricted segments such as the satellite.

Furthermore, there are two key options for QoS architectures which need to be taken into account.

The first assumes that the QoS is guided directly by the applications via an API or a protocol. The applications thus negotiate their QoS requirement with the network according to the services used which are the closest to their requirement. Nevertheless, this assumes that the application has the necessary information and the means of communicating this information, which remains relatively uncommon.

The second option consists of entrusting the QoS to the network without an explicit demand of the applications (only a marking or a network configuration are used in this case). This option allows applications to be accepted, without the need for modification for the QoS management, at the expense of a more refined QoS management (little useful information is available).

3.6.3.3. *Multimedia applications*

Since the BSM QoS architecture addresses a wide variety of applications, different types of multimedia services are possible, notably voice, audio, video, image and data services.

Figure 3.20 provides the classifications in terms of QoS of multimedia services, where different applications use components implementing the services, which themselves rely on various protocols, codecs and functionalities. The components are associated with QoS classes.

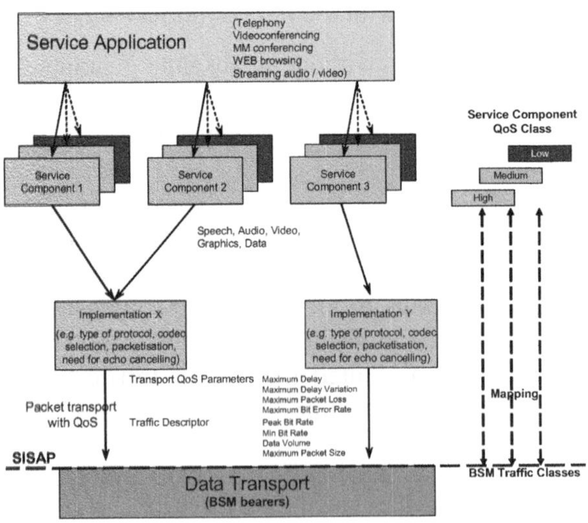

Figure 3.20. *Application and QoS framework*

The request for services can be either linked to the network, whereby the application contacts the operator's service control function (SCF) which is responsible for the implementation of the end-to-end service and associated QoS, or linked to the application, which manages the session itself as well as the QoS via a session/QoS protocol such as SIP/SDP.

3.6.3.4. *QoS scenarios*

Three scenarios are described for the negotiation of QoS in [ETS 06]:

– an *SP-oriented model.* In this scenario, the service is required with the SCF of the provider, which is responsible for determining the QoS requirements for the service and for ensuring that these requirements have been met with the network (or more precisely, with one or more resource controllers). Once permission is granted, the RC configures the various access and network components. This scenario helps to prevent reservation signaling in the user terminal, but means systematically using the SCF for any service demand;

– a *user-oriented model.* In this case, the user is able to indicate to the network operator his requirements in terms of QoS, without needing the SCF. The main restriction is therefore to use applications able to communicate their QoS requirements. However, this case represents a relatively significant separation between the network operator and the service operator, since the QoS reservations are made hop by hop and are not consolidated via a central element such as the SCF;

– an *application-oriented model.* This scenario combines the previous two approaches, with a user able to signal his QoS requirements (user-oriented model), which must be authorized before being able to use the service (SP-oriented model).

3.6.3.5. *BSM QoS architecture*

The general approach which is commonly accepted in IP networks is to separate the part related to the services from the part related to the network and communicating between each other via an open interface (see Figure 3.21). This type of architecture therefore enables a separate evolution of services and underlying networks.

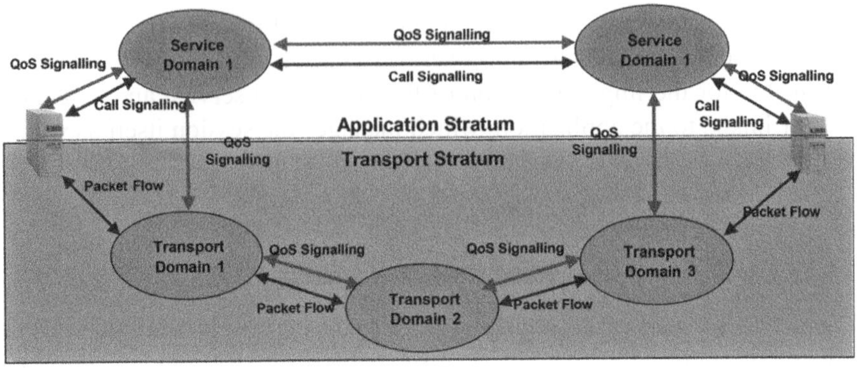

Figure 3.21. *General approach to QoS architectures*

The application part is responsible for the provision of services to users and uses a signaling protocol such as SIP. This signaling is particularly important because it describes the terminal points of the session (the client, or the caller and person being called), the QoS parameters (which depend on the protocol used), and also the parameters of the services themselves.

The transport part helps to transport the packets associated with the service and ensures the QoS level required by the application. To do this, a QoS signaling protocol is not only used (such as RSVP, COPS or NSIS), between the application part and the transport part, but also within the transport part between the various segments crossed.

It should be recalled that QoS is based on a set of blocks (as defined in [ITU 04]) which involve three planes – user, control and management (Figure 3.22). These blocks include the traditional functions of traffic classification (DiffServ or multi-fields), packet marking (used at the border for a given DiffServ domain), policing and/or shaping (with regard to a pre-established set of traffic) and scheduling (and queue management) for a *data plane.* The *control plane* features QoS routing functions (in order to select the pathway which respects the QoS requirements), along with service invocation, resource reservation and admission control (pathway based either on a measure a stream characteristics, or on their parameters if available). Finally, the *management plane* features functions of service subscription, SLAs, provisioning and billing.

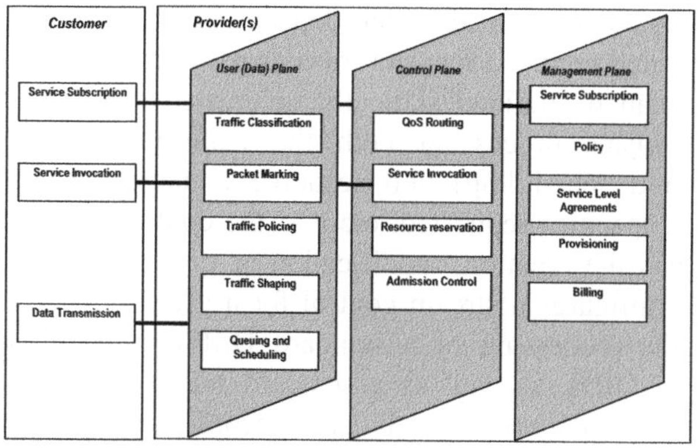

Figure 3.22. *Functional QoS architecture*

The BSM QoS is illustrated in Figure 3.23. In the control plane, it includes a SCF and element of resource management, in the satellite terminal and the gateway or NCC. These elements ensure the decision functions (PDP) and the implementation of QoS (PEP). These are connected both to the elements (client or server) of the high layers (proxy SIP, Agent RSVP/NSIS), and to the network and access layers of the data plane, in order to configure the various components responsible for QoS processing.

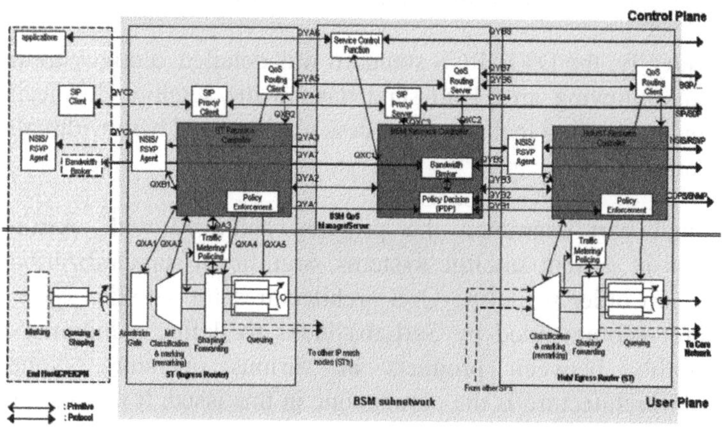

Figure 3.23. *BSM QoS architecture*

BSM introduces the notion of Queue ID (QID) to generically reference the MAC queues in SI-SAP. The QIDs are therefore an abstraction of the queues (the implementation of which depends of the various terminal/GW manufacturers) and are used for data transfer between the SI and SD layers. These queues can be defined statically or created dynamically, and they include mechanisms processing both the QoS at access level (queue management and scheduling) and resource management (types of CR, calculation of capacities requested, etc.) At the control plane level, SI-SAP not only enables the creation, modification or dropping of QIDs, but also activates and configures a stream control for a given traffic. This interface also enables the processing of a standard DiffServ model, with a static management of QIDs, as well as a standard IntServ model, including an admission control and a relatively dynamic reconfiguration of the parameters associated with the QID.

3.7. Conclusion

This chapter presented a state of the art of the DVB-S/S2 and DVB-RCS/RCS2 standards and the progression of the integration of IP and QoS into satellite networks.

First, the DVB-S standard was presented with various methods of IP encapsulation offering on the one hand by the DVB-S standard, and on the other hand by the IETF working group IP over DVB.

Subsequently, the DVB-RCS standard was detailed, along with the set of mechanisms allowing an ST to send data to the satellite. It specified the modulation and coding techniques, access methods and bandwidth allocation mechanisms.

The final part examined the progress achieved with existing QoS mechanisms in current satellite systems. Significant standardization efforts were made in order to define QoS architectures for satellite systems. The architecture recommended by SatLabs looks to define a minimal base of interoperability between products of various operators. Nonetheless, the *DiffServ* architecture is the cornerstone in this issue. It also examines the integration of satellite networks into wider *DiffServ* domains. Moreover, the functions of the control and management planes are specified in order to allow greater interoperability with next-generation networks.

The following chapter offers solutions to the integration of QoS in next-generation networks.

4

Integration of Satellites into IMS QoS Architecture

4.1. IMS architecture

One of the keys to the successful integration of a satellite system into terrestrial systems is the possibility of offering an end-to-end quality of service (QoS) guarantee. This point is essential since the satisfaction of end users and the guarantee of the correct use of network resources depend on it.

The internet protocol (IP) multimedia subsystem (IMS) architecture, initially developed by the third generation partnership project (3GPP) for universal mobile telecommunications system (UMTS) networks, offers an end-to-end QoS architecture which focuses on the access network part. This architecture aims to allow fixed-mobile convergence, which is the possibility to use a common infrastructure for wired and wireless networks. Its collaboration with ETSI-Telecommunications and Internet converged Services and Protocols for Advanced Networking (ETSI-TISPAN) has been a trigger to this. This chapter looks at present difficulties and opportunities to integrate satellite systems into a next-generation terrestrial infrastructure, such as the one proposed by IMS. This chapter will examine the traditional IMS architecture, as shown in Figure 4.1. The detail of the component elements is given in section 4.2.

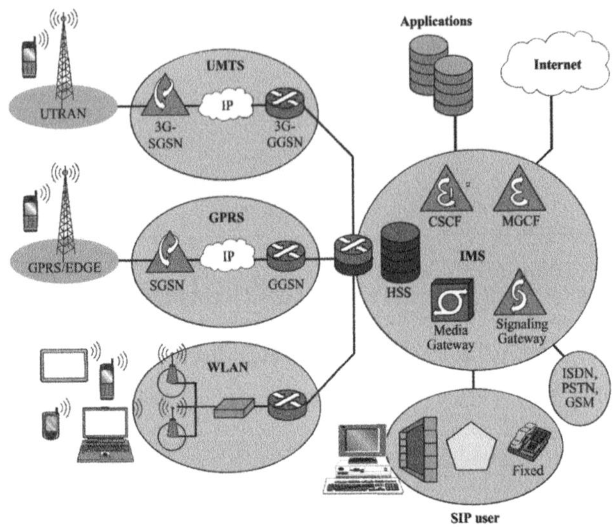

Figure 4.1. *Simplified IMS reference architecture*

Here, we consider horizontal integration as being the use of one or more access techniques connected to one or more IMS core networks.

There are a number of scenarios which may be considered:

– a single transparent satellite access segment (digital video broadcasting – return channel via satellite (DVB-RCS) standard) connected to an IMS core network;

– a single regenerative satellite access segment (meshed DVB-RCS standard) connected to an IMS core network;

– various access segments, including a satellite (DVB-RCS standard) connected to an IMS core network.

Moreover, horizontal integration covers both the end-to-end implementation of QoS between IMS users (e.g. during the set-up of a voice over IP (VoIP) session), and also the access to services provided by the network (notably, IMS services).

As will be demonstrated in the following sections, IMS chose, among others, Session Initiation Protocol (SIP) as a session control protocol, common open policy service (COPS) as a protocol for transferring policies and double RADIUS (DIAMETER) for authentication. These three protocols are, therefore, very important in the implementation of QoS. For ease of reading, we will repeat the messages exchanged by these protocols in the following section.

4.1.1. *COPS and DIAMETER messages*

4.1.1.1. *COPS*

The COPS protocol was presented in Chapter 2. For ease of reading the scenarios which use this protocol, we will briefly recall here its role and the messages it exchanges.

In order to improve the admission control network, in 2000, Internet engineering task force (IETF) defined an architecture based on the notion of policies. A policy is defined as a set of rules enabling the administration, management and control of the access to a network's resources. To allow the exchange of these policies in a client/server model, a protocol is also defined by this working group: the COPS policy.

This policy management model is made up of two central elements: the policy enforcement point (PEP), responsible for the application of policy decisions and the policy decision point (PDP), responsible for taking decisions based on the policy which was defined. These two elements communicate via the COPS protocol. Table 4.1 provides a list and the meaning of the COPS messages. For more details, readers can refer to [BOY 00].

Message PEP → PDP	*Message PDP → PEP*	*Message PEP ←→ PDP*
REQ : REQUEST (policy request)	*DEC* : DECISION (policy sending)	*CC* : CLIENT CLOSE (end of acceptance of COIPS client)
RPT : REPORT STATE (result of the policy implementation)	*SSQ* : SYNCHRONISATION STATE REQUEST (synchronization request)	*KA* : KEEP ALIVE (existence signal)
DRQ : DELETE REQUEST STATE (end of the policy application)	*CAT* : CLIENT ACCEPT (accepting of COPS client)	
OPN : CLIENT OPEN (request to accept COPS client)		
SSC : SYNCHRONIZE STATE COMPLETE (end of synchronization)		

Table 4.1. *List and meaning of COPS messages*

4.1.1.2. *DIAMETER*

The DIAMETER protocol [CAL 03] is the evolution of the remote authentication dial-in user service (RADIUS) protocol [RIG 00a, RIG 00b] with a number of improvements. It is generally considered as being the reference of Authentication, Authorization, Accounting (AAA) protocols. This protocol was very widely used in long-term evolution (LTE) and IMS architectures to exchange AAA information.

The DIAMETER architecture is made up of several entities which are nodes, clients, servers and agents:

– A DIAMETER node is a host which implements the DIAMETER protocol;

– A DIAMETER client is a node on the border of the network which carries out access control. Examples of DIAMETER clients are serving GPRS support nodes (SGSNs);

– A DIAMETER server is responsible for requests for authentication, authorization and taxation for a given domain;

– A DIAMETER agent is a DIAMETER node which provides gateway (GW), proxy or translation services. An example of a proxy agent is the policy and charging rules function (PCRF) entity in the LTE architecture. An example of a redirection agent is the subscriber location function (SLF) entity in the IMS architecture.

Table 4.2 provides a list of the main messages defined by the DIAMETER protocol. For more details, refer to [BOY 00].

4.2. IMS QoS architecture

The implementation of QoS in an IMS architecture depends on various functional entities:

– the call session control function (CSCF) is the standardized access point of the IMS user. It manages the authentication, access to services and the service policy. According to the network architecture itself, the CSCF may be a proxy (P-CSCF), a server (S-CSCF) or a border element (I-CSCF);

– the policy decision function (PDF) is responsible for decisions in terms of service policy;

– policy enforcement functions (PEFs) ensure the regulation of IP datagrams (traffic conditioner) and the filtering of possible destinations (classifier);

– application functions (AFs) offer services to IMS users.

Message name	Abbreviation	Command code
AA-Request	AAR	265
AA-Answer	AAA	265
4.1.1.2.1. *Abort-Session-Request*	ASR	274
4.1.1.2.2. *Abort-Session-Answer*	ASA	274
4.1.1.2.3. *Accounting-Request*	ACR	271
4.1.1.2.4. *Accounting-Answer*	ACA	271
4.1.1.2.5. *Capabilities-Exchanging-Request*	CER	257
4.1.1.2.6. *Capabilities-Exchanging-Answer*	CEA	257
4.1.1.2.7. *Device-Watchdog-Request*	DWR	280
4.1.1.2.8. *Device-Watchdog-Answer*	DWA	280
4.1.1.2.9. *Disconnect-Peer-Request*	DPR	282
4.1.1.2.10. *Disconnect-Peer-Answer*	DPA	282
4.1.1.2.11. *Re-Auth-Request*	RAR	258
4.1.1.2.12. *Re-Auth-Answer*	RAA	258
4.1.1.2.13. *Session-Termination-Request*	STR	275
4.1.1.2.14. Session-Termination-Answer	STA	275

Table 4.2. *List of messages defined in the DIAMETER protocol*

It should be noted that the PDF may be an element of the CSCF, or a distinct node connected to it via a standardized interface. More precisely, using version 5 of IMS, the interface between the P-CSCF and PDF was opened, enabling a separation of the two elements. Subsequently, version 6 defined the interface between them (DIAMETER). The role of PEF can be ensured in the gateway GPRS support node (GGSN), at the interconnection with the border network. Figure 4.2 presents this architecture.

These elements are clearly connected to the border equipment of the access segments linked to the IMS network. Therefore, in a UMTS network, the PDF has an interface with the GGSN that ensures the interconnection between the general packet radio service (2.5G) (GPRS) network or UMTS on one side and the IP networks (here, IMS) on the other side.

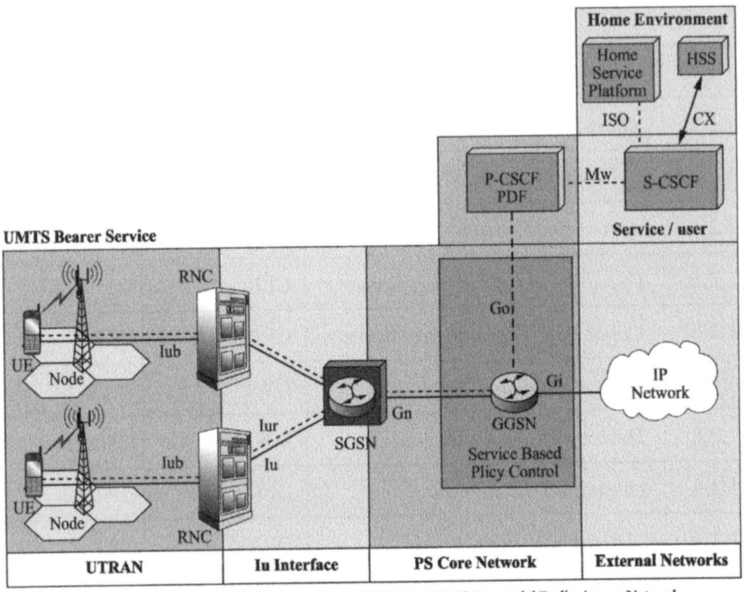

ISO : Intertional Standards Organizatioin UTRAN : UMTS Terrestrial Radio Access Network
RNC : Radio Network Controller

Figure 4.2. *IMS architecture*

In general, IMS does not directly implement QoS but relies on the underlying networks which must be able to guarantee a certain level of QoS, depending on the client services. However, IMS offers a unified and common interface for managing QoS.

It is possible to distinguish the two distinct domains for the management of QoS in IMS:

– the access network: QoS management solutions are dependent on the underlying access technology and the equipment used. Nevertheless, two interfaces are implemented here. The first uses COPS for communication between the PDF and the access segment, and the second is based on the sending of SIP messages between the end user terminal (or terminal node) and the P-CSCF. It should be noted finally that a PDP context is used between the terminal node user equipment (UE) and the access segment;

– the core network: QoS management solutions rely on the traditional IP mechanisms, notably DiffServ and IntServ/Resource Reservation Protocol (RSVP) [ETS 11]. IMS does not define the actual management method of the QoS in the core network and relies on provision mechanisms. However, the inclusion of an IntServ model in IMS can provide configuration solutions in the core network via RSVP.

4.2.1. *IMS QoS in a GPRS: UMTS network*

In order to gain a clear understanding of the QoS architecture defined in IMS, we will specifically examine its reference implementation, notably that offered by a GPRS or UMTS network (Figure 4.3).

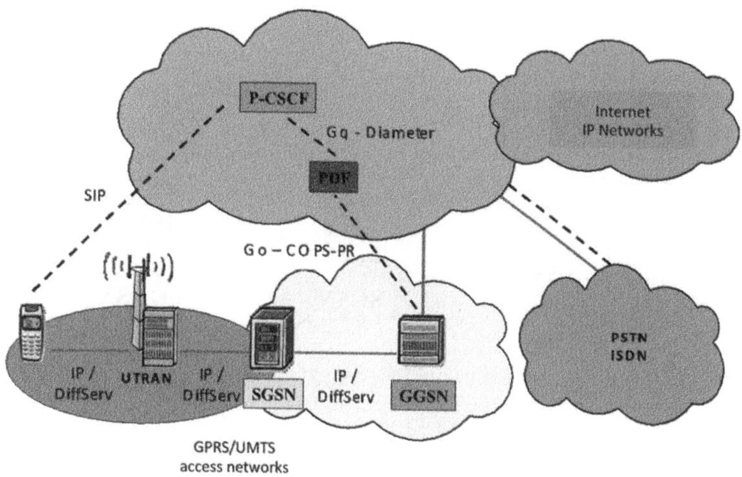

Figure 4.3. *IMS UMTS QoS architecture*

Figure 4.4 shows an example of an opening of an IMS session in a GPRS/UMTS framework. Figure 4.5 gives an example of PDP context in a GPRS/UMTS network.

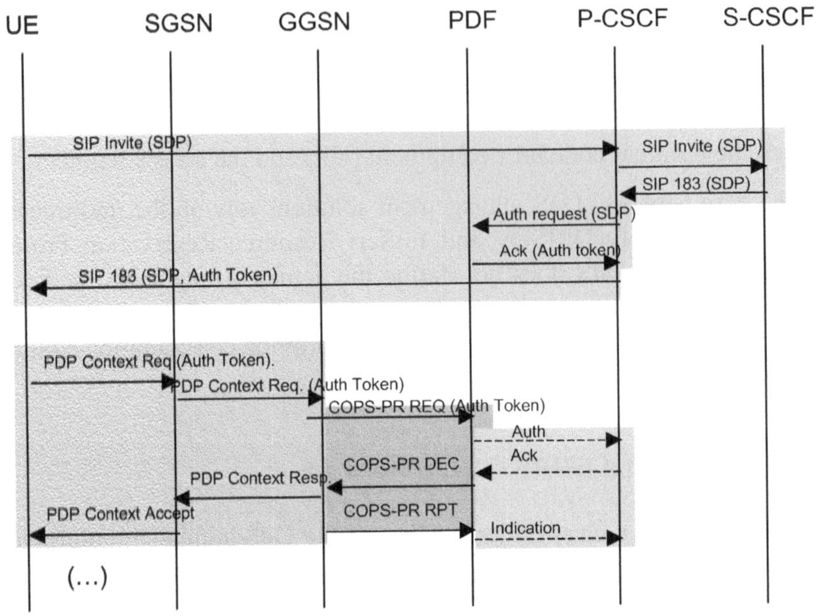

Figure 4.4. *Example of an opening of an IMS session*

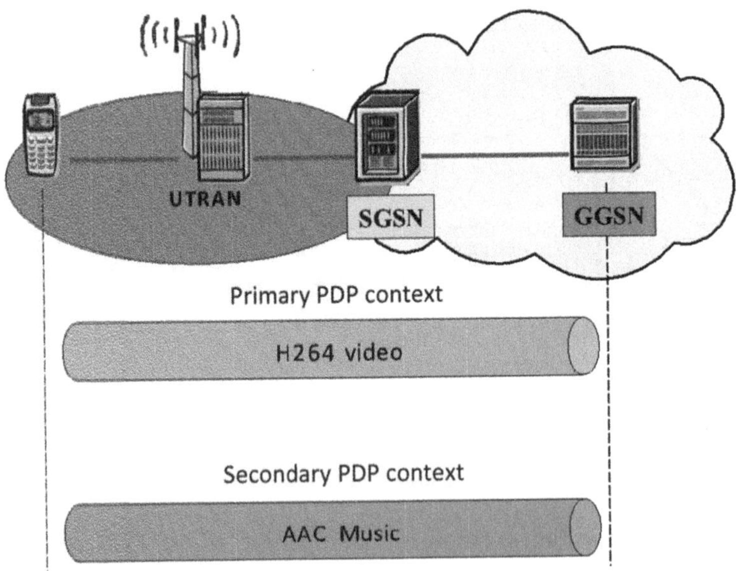

Figure 4.5. *PDP context in a GPRS UMTS network*

As can be observed (Figure 4.4), three phases can be distinguished:

1) The initiation of the IMS session, which is traditionally carried out via the SIP protocol. Thus, the user terminal indicates its QoS requirements (via Session Description Protocol (SDP)) to its proxy CSCF, which then relays it to the S-CSCF, and finally to the distant terminal. This exchange also involves the PDF, which authorizes or refuses the session depending on the required parameters. It issues an authorization token which is used later on.

2) The initialization of the PDP context, between the terminal node and the SGSN, and then the GGSN (which plays the role of PEF here). These contexts include the addresses used as well as the QoS parameters. ETSI TS 23.107 [ETS 12b] defines the set of possible parameters, notably, the maximum transfer rate, the guarantee of provision of packets in order, the maximum size of the SDU, the list of service data unit (SDU) sizes authorized, the SDU error rate, the binary residual error rate and the provision of incorrect SDUs.

3) The implementation of the QoS requested by the GGSN to the PDF via a common open policy service – policy provisioning (COPS-PR) request. The PDF, therefore, plays the role of COPS server, and thus of PDP.

As well as the GGSN, the QoS is implemented in compliance with the architecture of the core network. In the case of a DiffServ network, for example, the GGSN is responsible for positioning the differentiated services code points (DSCP) in compliance with the defined strategy. It can also use RSVP to implement this QoS. Here, the GGSN, therefore, adopts the role of domain border router.

In summary, two important points should be retained:

– first, the two generic interfaces are independent from the existing underlying access segment and therefore from the specific protocols used. On the one hand, the end-to-end signaling of the session is based on the SIP protocol and on the other hand, the signaling between the access network and the core IMS network depends on the COPS-PR (Go interface) and DIAMETER (Gq interface);

– QoS signaling within the access network is dependent on the technology. Therefore, in the framework of a GPRS/UMTS network, this signaling relies on PDP contexts which ensure the IP connectivity and the definition of QoS parameters.

4.2.2. *IMS QoS in an asymmetric digital subscriber line (ADSL) network*

ETSI TS 123.228 [ETS 08] also indicates the opening procedure of a session on an xDSL network (Figure 4.6 and 4.7). This definition, even though it does not constitute a normative framework, provides a clear illustration of the various modulations possible in IMS QoS architecture according to the specificities of the access segments.

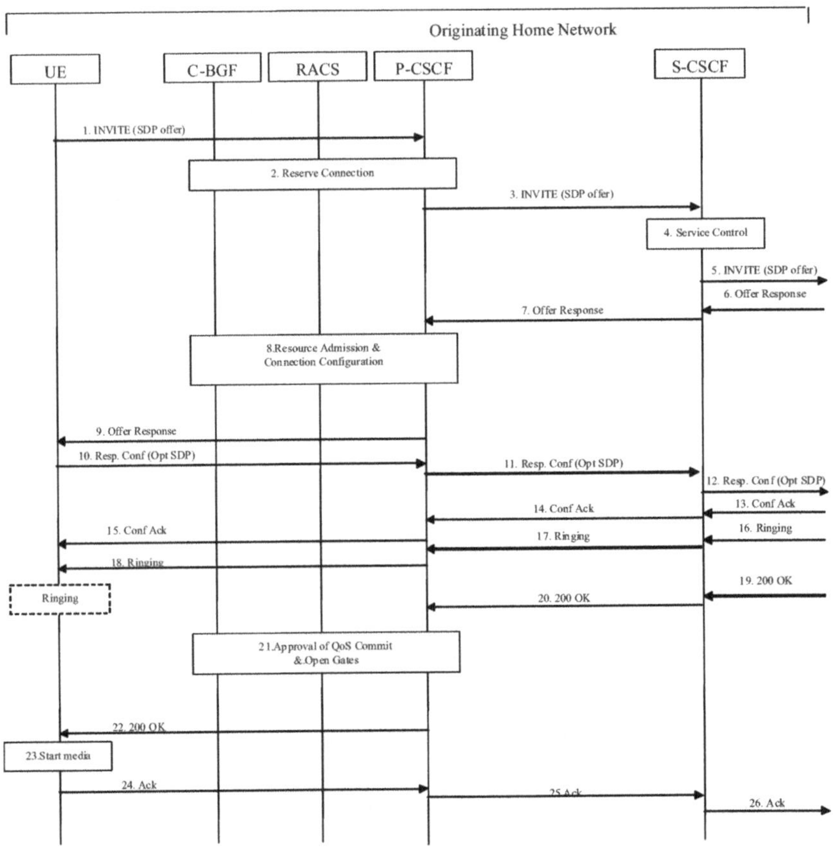

Figure 4.6. *Opening procedure of an IMS session in an xDSL network (source node side)*

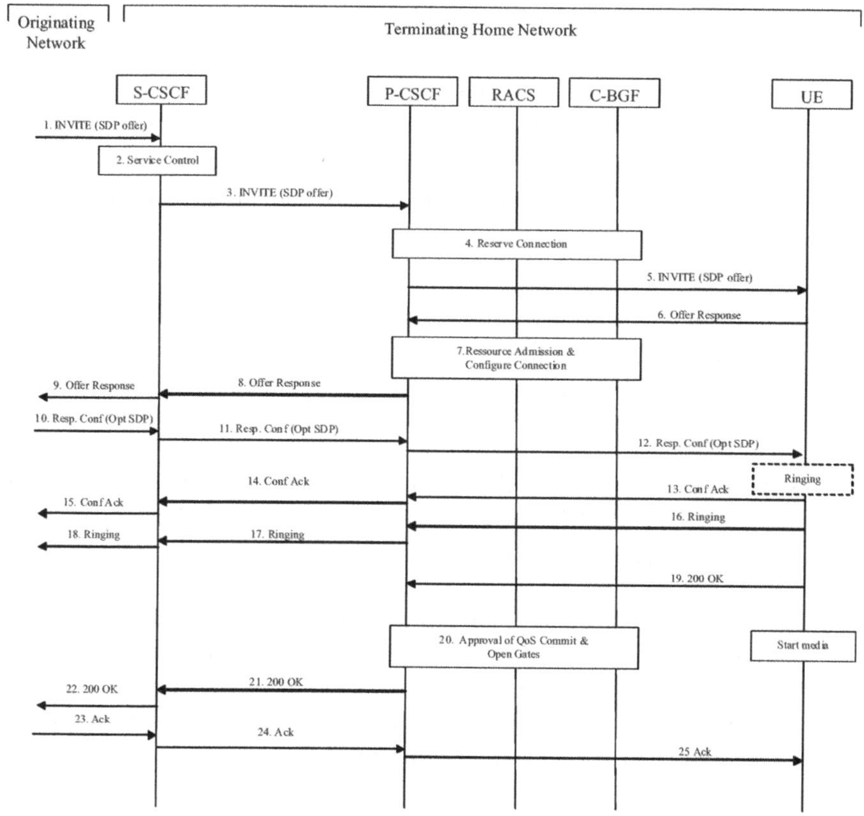

Figure 4.7. *Opening procedure of an IMS session in an xDSL network (destination node side)*

Therefore, the first phase is generally equivalent to the one encountered in a UMTS network (although with several differences related to the network address translation (NAT)) since it is based on SIP, and the configuration of QoS parameters at the level of the access segment (phases 2 and 3 previously) is directly associated with the SIP protocol (upon receiving the offer response and the SIP OK). In particular, it is not initiated by the terminal node (UE) as previously in the network.

4.3. IMS QoS signaling

Two cases can be distinguished for the procedures enabling interaction between QoS and IMS signaling:

– when the IP connectivity transporter is activated, the user can access services according to the choice of operator with a local service policy mechanism. An IMS service policy function ensured by the PEF is then used. This functionality regulates the IP packets and reduces the accessible destinations (defined with a classifier), thus defining an open "gate". Control is carried out by a PDF which may be an entity of the proxy control server or on a separate node. In this case, the functions follow a standardized interface with the proxy;

– when the IP connectivity transporter is deactivated, the transporter uses the subscription characteristics and the admission controls of the local transporter. Since it is not integrated by its very nature, this case will not be examined in this study.

IMS defines a set of procedures which enable the management of the QoS. The signaling used in the following sections uses as a hypothesis a GPRS or UMTS access segment.

It is also worth noting that ETSI TS 29.208 [ETS 07b] addresses general cases but provides almost no descriptions of degraded cases (only the case of the GGSN receiving an unsolicited authorization decision from the PDF can be considered as non-nominal operation.) Finally, IMS does not specify the overall resource management algorithms implemented to accept or refuse sessions.

4.3.1. *Authorization of QoS resources*

The procedure authorizing the reservation of QoS resources is carried out during the establishment or modification of a SIP session. The P-CSCF proxy uses SDP information contained in SIP signaling to apply the session information necessary for the local service policy and relay them to the PDF. This allows the PDF to authorize or refuse the use of resources (admission control).

Figure 4.8 shows the procedure at the level of the source PDF.

Figure 4.9 describes the same procedure on the destination side, for a session initiated by an Internet node, for example, or in the case of a session between two nodes connected to access networks which are themselves attached to an IMS network.

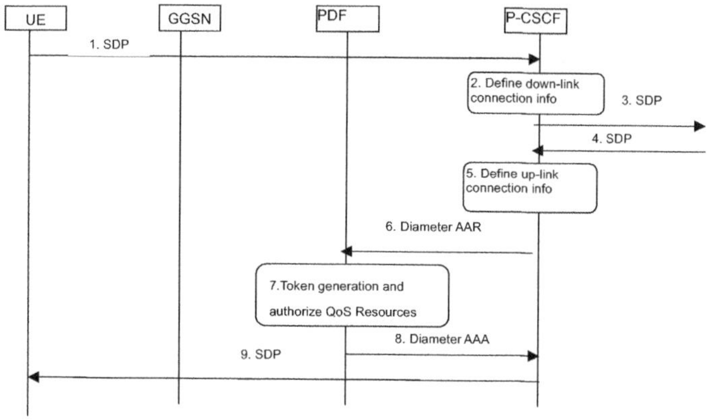

Figure 4.8. *QoS resource authorization procedure in the source PDF*

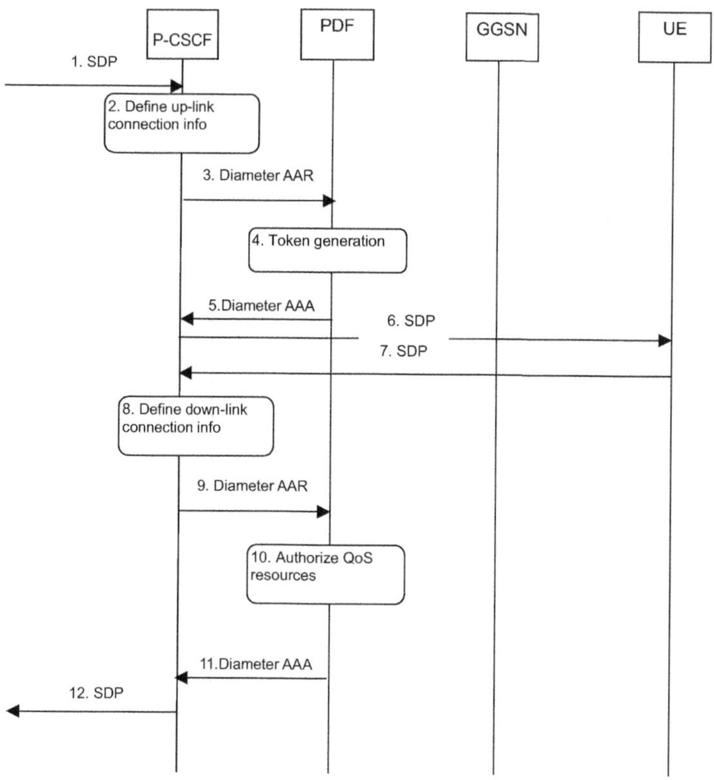

Figure 4.9. *QoS resource authorization procedure in the destination PDF*

In this case, it can be noted that, unlike the first case, the authorization (and therefore the token generation) is separated from the QoS resource authorization.

In the event of a modification of the session and therefore of the resources associated with it, a pair of SIP Re-INVITE/OK messages is exchanged by the border entities (UE). As in the previous two cases (Figure 4.8 and 4.9), the P-CSCF intercepts the request and the response, identifies the changes to make to the session using the SDP content and the session history and then makes an authorization request to the PDF via a DIAMETER AAR request. This procedure is valid for both terminals.

The authorization of a session given by the PDF can lead to the approval (section 4.3.3) or deletion (section 4.3.4) of the QoS commitment, the revocation of authorization following the deletion of a media component (section 4.3.5) or finally, a session modification in order to update service information (section 4.3.7).

4.3.2. *Reservation of QoS resources with a local service policy*

This procedure enables the implementation of a QoS policy at the PEF level, which is in the GGSN in a UMTS architecture. The reservation decision will be taken by the PDF according to the parameters of the QoS resource request. An optional AF authorization may be necessary (i.e. of the service applied).

The scenario is initiated by the UE at the request of a PDP context. The SGSN relays the request to the GGSN which makes an authorization request to the PDF. The PDF may verify the identity of the UE with the application (AF), and then confirm its decision with the GGSN. The GGSN will transmit to the PDF the result of the policy decision at the same time as it implements it in order to create a PDP context. The PDF may optionally inform the application.

4.3.3. *Approval of commitments of authorized resources*

When QoS resources are not activated at the moment of reservation on the basis of a local service policy, or if the stream or streams previously paused is or are reactivated (via SDP direction messages: sendrecv, sendonly, recvonly or no direction), this procedure allows the PDF to take the decision to use these resources and to inform the PEF (GGSN) that the user is now able to use the resources authorized for this session.

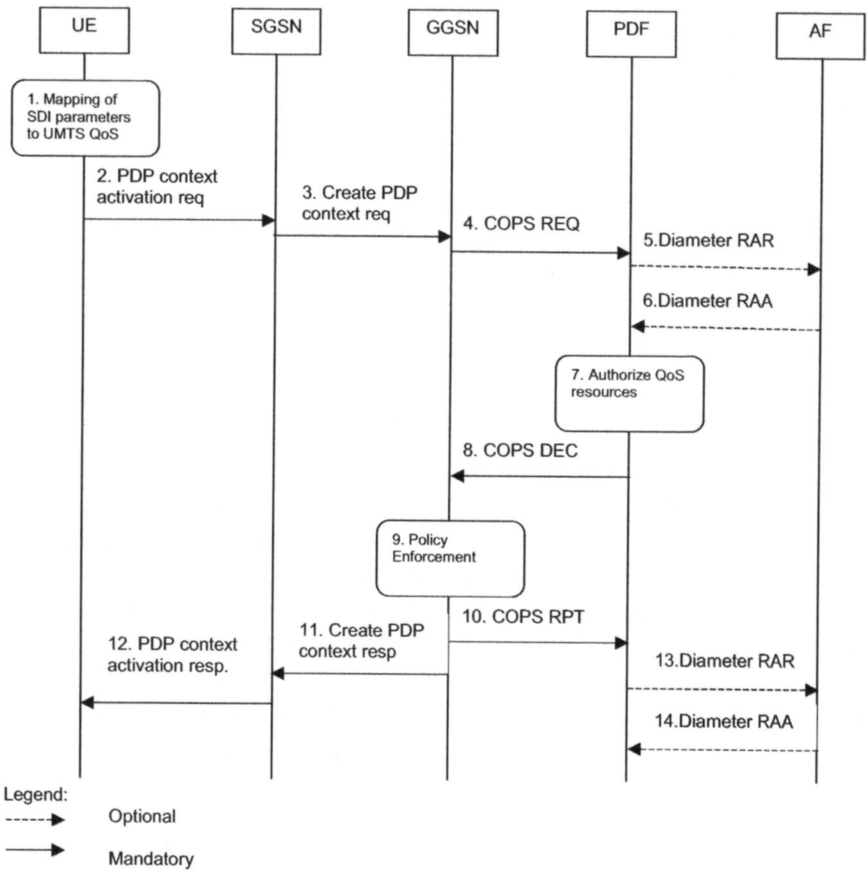

Figure 4.10. *Resource reservation procedure with a local service policy*

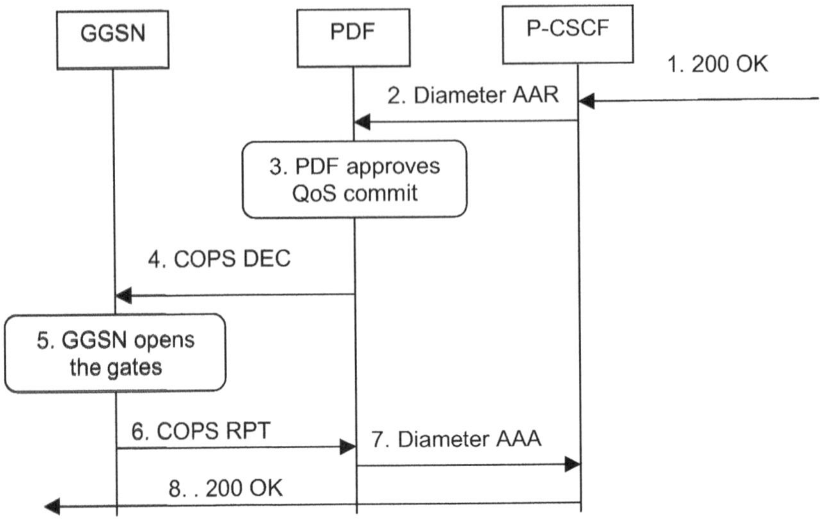

Figure 4.11. *Procedure for the approval of commitments of authorized resources*

4.3.4. *Deleting commitments of authorized resources*

This procedure is equivalent to the previous procedure and is used to manage the pause of a media or the deletion of a media making up part of the session. During the interception of an SDP request for a pause or deletion of a media, like in the previous case, the P-CSCF relays the request to the user terminal, and then informs the PDF via DIAMETER (AAR), which notifies the GGSN of the policy change (COPS DEC). The confirmation from the GGSN passes to the PDF via COPS (RPT) and to the P-CSCF via DIAMETER (AAA). It is advised to initiate a timer in the PDF to ensure this change takes place.

The procedure of deleting a QoS commitment following the deletion of a media in the session is optional.

4.3.5. *Revocation of a QoS resource authorization*

The procedure of revoking the authorization is implemented following the closure or redirection of the last (or only) session of a user. It can also be activated in the case of a negative response to the initial "bearer establishment".

Figure 4.12 describes the procedure of a revocation initiated by the mobile node. It is valid for both the source and destination.

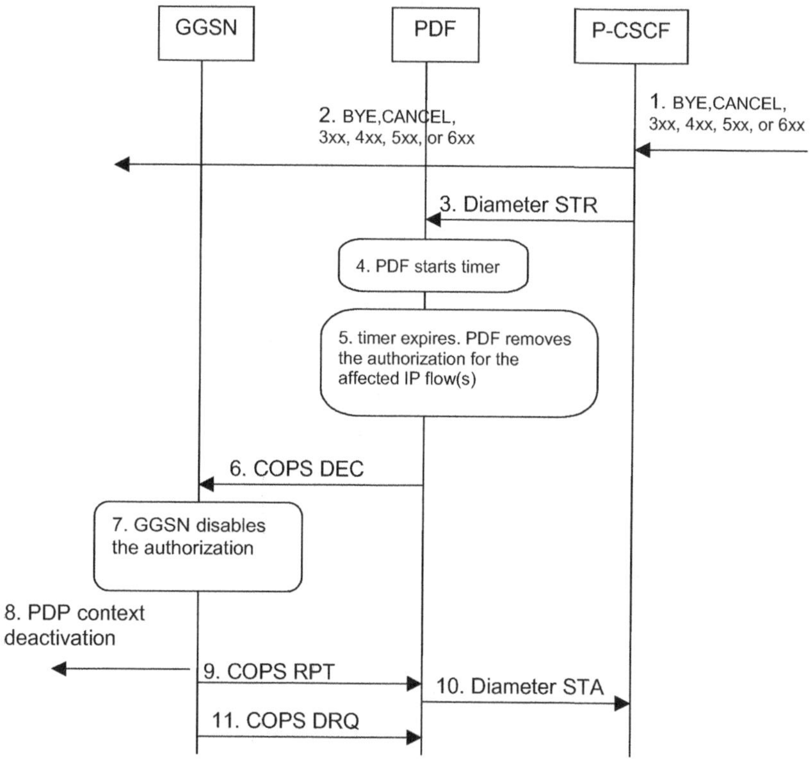

Figure 4.12. *Procedure of revoking authorization initiated by a mobile or network node*

4.3.6. *Indication of a PDP context deletion*

This procedure is implemented during the closure of the PDP context. This indication can be initiated normally by the SGSN (Figure 4.13), and also by the GGSN in the case of an error or by the PDF in the case of an error or a network overload. The last two cases will be presented subsequently.

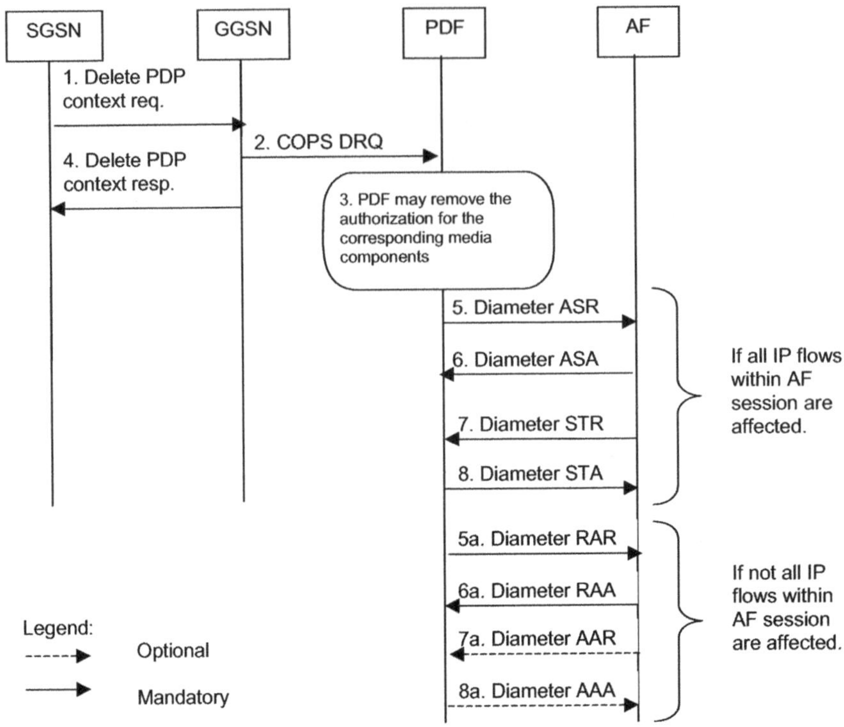

Figure 4.13. *Indication of PDP context deletion*

Upon receiving a request for the deletion of the policy (DRQ), the PDF notifies the application of the end of the session (ASR), and then confirms the termination of the session (STA) if all the application streams are affected. If this is not the case, a modification will take place (RAR).

If the revocation is initiated by the GGSN, it will request the deletion of the PDP context to the SGSN, and then will inform the PDF of the end of the policy application (DRQ). The PDF will inform the application in the same way as previously via DIAMETER.

If the revocation is initiated by the PDF, the decision is transferred to the GGSN via a COPS (DEC) message, which will request the SGSN for the deletion of the context. At the end of the procedure, the GGSN will confirm the deletion with the PDF via (RPT) and (DRQ) messages.

The PDF will inform the application in the same way as previously via DIAMETER.

4.3.7. *Authorization for the modification of the PDP context*

This procedure is activated when the PDP context is modified so that the QoS requested exceeds the limits authorized by the activation of the PDP context or the maximum forward and return transfer rate is 0 kbits/s.

Two mechanisms are implemented:

– procedure for authorizing the modification of the PDP context;

– procedure for indicating the modification of the PDP context.

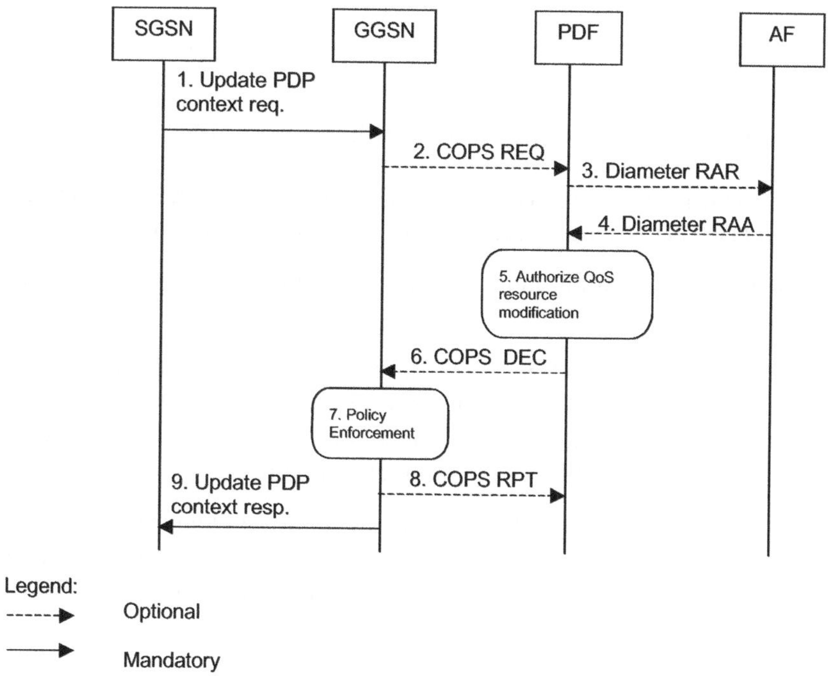

Figure 4.14. *Authorization procedure for the modification of the PDP context*

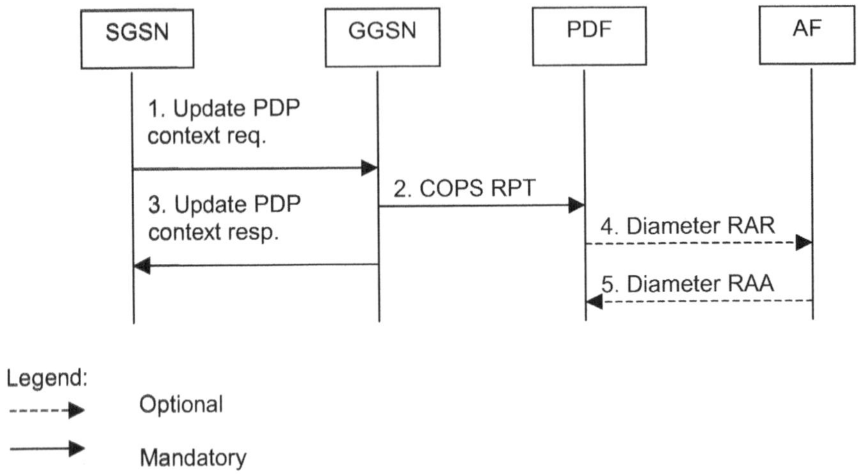

Figure 4.15. *Indication procedure for the modification of the PDP context*

4.4. Inclusion of IMS QoS in the satellite segment

As shown in the previous section, the horizontal integration of QoS in a standard DVB-RCS satellite architecture must satisfy a number of demands:

– first, its integration must be based on the end-to-end SIP session signaling;

– subsequently, the QoS architecture must be based on the elements inherent to IMS on one hand (mainly, P-CSCF and PDF) and on QoS components specific to the access segment on the other hand. Above all, it is a question of implementing QoS "enforcement points";

– the satellite segment must comply with the various QoS procedures defined, if they are relevant to this segment. In particular, we will examine the procedures linked to the PDP contexts which are specific to the GPRS/UMTS network.

4.4.1. *"System" hypothesis*

In order to determine the adaptations necessary for the definition of QoS architecture which is compatible with IMS, it is necessary to identify the various hypotheses of the satellite segment:

– the first hypothesis regards the positioning of the various IMS elements in the network. From a general point of view, the P-CSCF may be colocated with the satellite terminal (ST) or with the GW depending on scenarios (transparent or integrated approach). Likewise, the PEFs also need to be positioned in this type of architecture;

– the second hypothesis regards the mode of IP transport within the access network and the UMTS segment. Therefore, the use of a PDP context can be integrated into a DVB-RCS system with an access connection, possibly established due to a C2P protocol [ETS 09f]. In this framework, the different procedures relating to the PDP contexts could be adapted taking account of the implementation of specific access connections with the segment. If this is not the case, for example if C2P connections cannot be made to the end user (if the ST is not the end user), an ADSL modulation could be used. Indeed, IMS QoS procedures defined for xDSL networks are initiated by the network and not by the user;

– the third hypothesis is the consequence of the previous one and regards the choice of QoS mechanisms and protocols within the access segment. Thus, the use of COPS between the PDF and PEF needs validating;

– the final hypothesis relates to the QoS model to be implemented in the access network. Here, IMS remains relatively flexible and does not advocate any particular solution. DiffServ and IntServ are, therefore, two possible options in an IP network. Nevertheless, the problems inherent to IntServ, notably the complexity, the scaling-up and the need for architecture and elements which are compatible with the entire pathway to ensure end-to-end QoS, clearly demonstrate the preference for a DiffServ model. This model will, therefore, be used in the remainder of the chapter.

The inclusion of QoS in the satellite segment depends on the different scenarios in question. These scenarios are described in the following section.

4.4.2. *IMS satellite integration: transparent approach*

The first architecture examined is based on the distribution of elements enabling completely transparent integration of a satellite segment into IMS. Therefore, the P-CSCF is colocated with the GW, and no prior modification or adaptation of the architecture of the satellite telecommunication system is necessary. Of course, the QoS functionalities can be linked via a traditional

PDP/PEP architecture, since PEPs are the responsibility of the satellite segment. Moreover, this architecture addresses the IMS client STs.

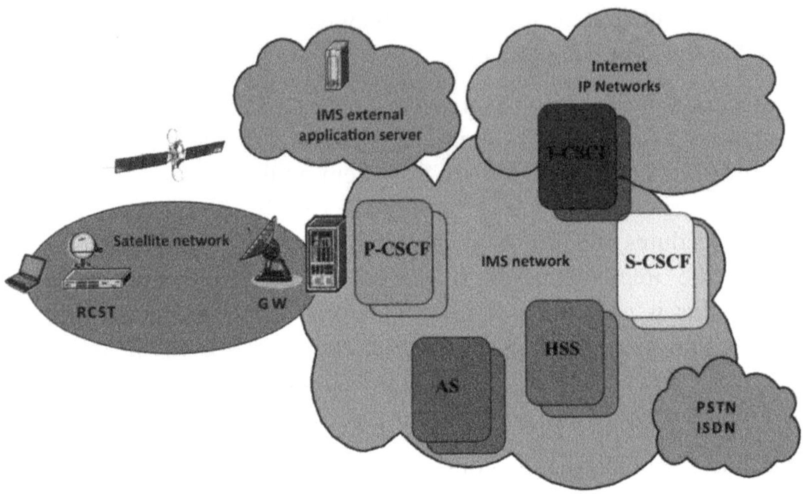

Figure 4.16. *IMS architecture – satellite – transparent integration*

In this type of architecture, the PDP is naturally colocated with the P-CSCF. There are two options for the location of the elements implementing the QoS policy:

1) a PEP in the terminal for the return channel and a PEP in the GW for the forward channel;

2) a PEP in the GW for both directions. The inclusion of QoS in the return channel is thus the responsibility of the access layer, or the local policy of the satellite segment.

From an end-to-end point of view, solution 1 will be prioritized. It can also be noted that other control elements can be added, for example a PEP in a border router if it is different from the GW.

4.4.3. *IMS satellite integration: integrated star approach*

The second architecture examined includes an IMS element in the satellite system, namely the P-CSCF which is colocated with the ST.

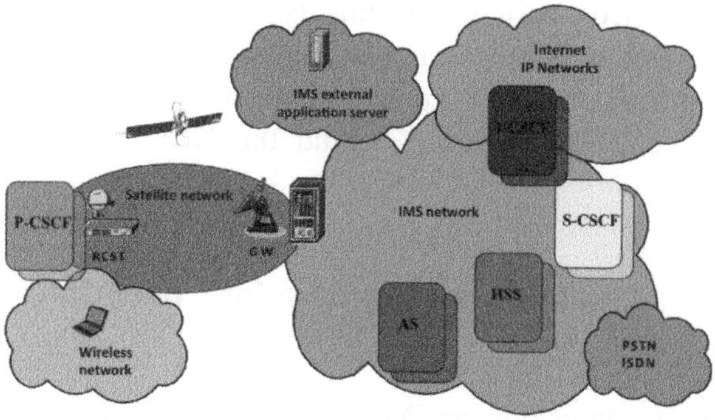

Figure 4.17. *IMS architecture – satellite – integrated star approach*

For this architecture, the PDP is also positioned behind the GW. Logically, two PEPs are used behind the GW and behind the ST, respectively.

4.4.4. *IMS satellite integration: integrated mesh approach*

The last architecture examined is based on a satellite system which offers meshed connectivity, whether created with a transparent satellite, a satellite with a switching matrix or an in-built IP router. This type of system enables the management of various IMS networks, each one made up of P-CSCFs colocated with STs and an S-CSCF (and other core IMS elements) behind a terminal providing access to a home network of the IMS network. It will also be noted that in this architecture, the I-CSCF can enable the interconnection with terrestrial IMS networks, and also with other IMS networks managed by the satellite network.

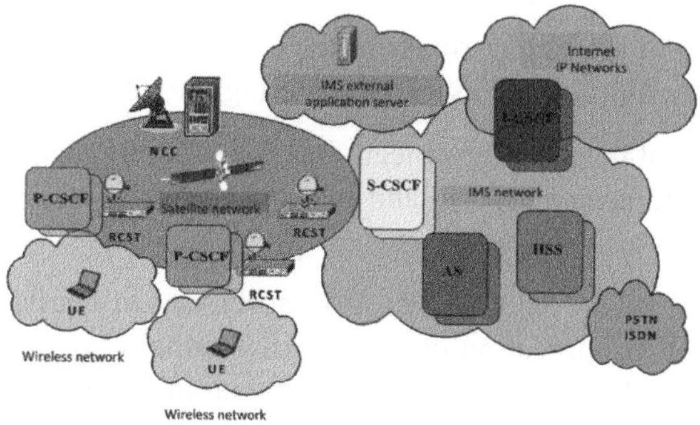

Figure 4.18. *IMS architecture – satellite – integrated mesh approach*

Given the meshed topology of this type of system, the C2P access connection protocol will be used. For this architecture, the PDP remains positioned behind the network control center (NCC). Unlike the previous architecture, three PEPs are used behind the NCC and behind the STs, respectively.

4.5. Toward a unified next-generation network (NGN) QoS architecture

This section examines how an IMS QoS architecture can be deployed in a satellite context. The IMS scenarios presented at the beginning of this chapter will be adjusted according to the approaches chosen for the integration of the satellite in the terrestrial network.

4.5.1. *Transparent integration scenario*

4.5.1.1. *Architecture*

In the architecture developed, end-to-end signaling is ensured by the SIP session protocol. At the level of the access segments, particularly the satellite access network, the P-CSCF contacts the PDF. It is, therefore, responsible for the acceptance of the connection (session) according to the information available at the level of the NCC of the satellite network. After verifying that the resources are sufficient for ensuring the QoS level required for the establishment of the session, the PDF contacts the various PEFs so that they implement the corresponding QoS policy.

Different PEFs can be proposed:

– at the level of the border router of the access network. In essence, it is a question of controlling the transfer rate entering into the Satcom network;

– at the level of the ST. The objective is thus to configure the transfer rate entering from the user network, and to open the gates associated with the service (IP addresses, port, etc.);

– at the level of the NCC. Here, it is necessary to configure the access parameters for communication on the return channel (e.g. by modifying the parameters of the allocation loop for the terminal in question);

– at the level of the GW. In this case, the PEP configures the scheduler for the traffic on the forward channel.

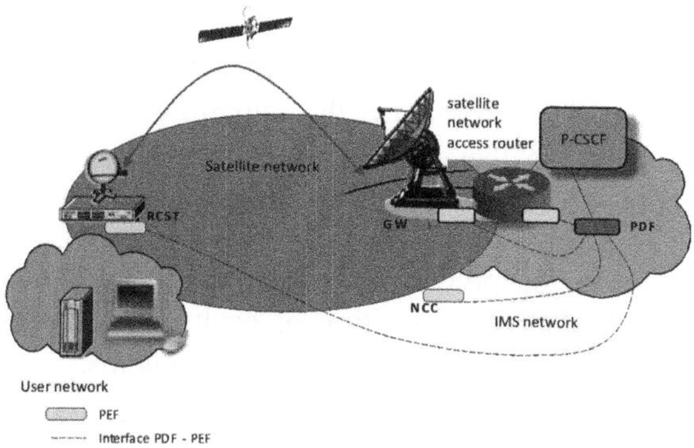

Figure 4.19. *IMS satellite architecture in scenario 1*

4.5.1.2. *Signaling*

4.5.1.2.1. Signaling at the network and session level

Figure 4.20 describes the signaling necessary to implement the QoS during the establishment of an IMS session. This is based on an xDSL-oriented approach, where the network is responsible for the implementation of the QoS.

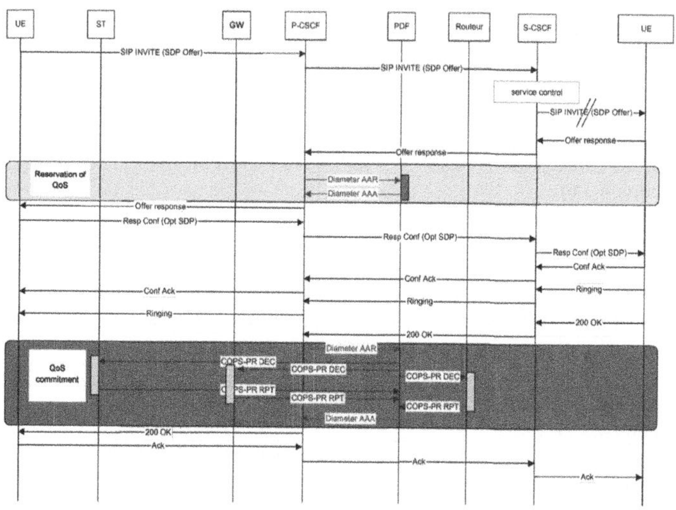

Figure 4.20. *General implementation of QoS for transparent integration*

The impact of this signaling is mainly located at the level of the implementation of QoS policies, which require two extra satellite hops for the routing of COPS-PR messages.

It appears that the set of procedures defined in [ETS 07b] cannot be applied in the satellite context. Table 4.3 shows the suitability of IMS QoS procedures in the satellite context, in the absence of equivalence with PDP contexts. Nevertheless, section 4.5.1.2.2 offers a signaling solution to this problem.

Column 1 refers to the procedures presented in section 4.3, column 2 indicates the suitability for transposition of each procedure onto a satellite network and column 3 lists any inherent-related problems.

Procedure	*V*	*Comments*
Authorization of QoS resources (section 4.3.1)	✓	
Authorization of QoS resources during a modification	✓	
Reservation of resources with a local service policy (section 4.3.2)	✗	In the absence of an access connection (and thus correspondence with PDP contexts), this procedure is not implemented in a satellite context
Approval of commitments of authorized resources (section 4.3.3)	✓	More PEFs are implemented in the case of satellites than for GPRS/UMTS
Deletion of commitments of authorized resources (media pause) (section 4.3.4)	✓	
Deletion of commitments of authorized resources (media deletion) (section 4.3.4)	✓	
Revocation of authorization initiated by the mobile or network node (section 4.3.5)	✓	
Revocation of authorization in the case of the deletion of a multimedia component (section 4.3.5)	✓	The deactivation of the PDP context would not be useful in this case
Indication of the deletion of the PDP context (section 4.3.6)	✗	No PDP context
Indication of the deletion of the PDP context initiated by the GGSN (section 4.3.6)	✗	No PDP context
Indication of the deletion of the PDP context initiated by the PDF (section 4.3.6)	✗	Only the indication initiated by the PDF may be used in the case of network overload
Authorization of modification of the PDP context (section 4.3.7)	✗	No PDP context
Indication of modification of the PDP context (section 4.3.7)	✗	No PDP context
Authorization of local service policy initiated by a session modification	✗	Unsolicited implementation of a session modification. No PDP context

Table 4.3. *Suitability of IMS QoS procedures in a satellite context*

4.5.1.2.2. Access network and session-level signaling

The implementation of the access connection within the satellite segment helps to extend the horizontal integration achieved at the higher levels. C2P [ETS 09f, ETS 09e] specifies the mechanisms which take account of the QoS at medium access control (MAC) level, with regard to the network QoS models. This approach is particularly interesting in the meshed case (since it is necessary to establish this type of connections between the terminals). In the framework of IMS architecture, the use of C2P also helps to imitate the operation of GPRS and UMTS network based on PDP contexts, for IP traffic. However, there are a number of problems to tackle:

– on the one hand, the access connections do not reach the end user (UE): they are open between the ST and GW or between the STs, since the NCC is always involved in the signaling. Thus, opening the access connection with the appropriate QoS parameters means that either the user is able to provide this information (from the SDP) to the ST via any type of protocol (SIP, simple object access protocol (SOAP) and web services), or that the establishment of the C2P connection is initiated directly by the NCC;

– on the other hand, another problem is raised by the delay and the band used for the routing of the C2P signaling. This is particularly considered in a transparent architecture which does not require prior connection at access level;

– finally, the relevance of the duplication of information between the different levels constitutes the major issue in this reflection. Indeed, C2P and COPS are clearly used in similar roles. C2P generates the gates at the access level, while COPSs do likewise at the network level. They will also be responsible for the QoS policy itself. This coverage at the operational level is even more important if we consider the cross-layer solutions between the session, network and access layers.

In order to highlight and compare the two solutions, the proposal for horizontal integration combining signaling at levels 5, 3 and 2 is described in Figures 4.21 and 4.22.

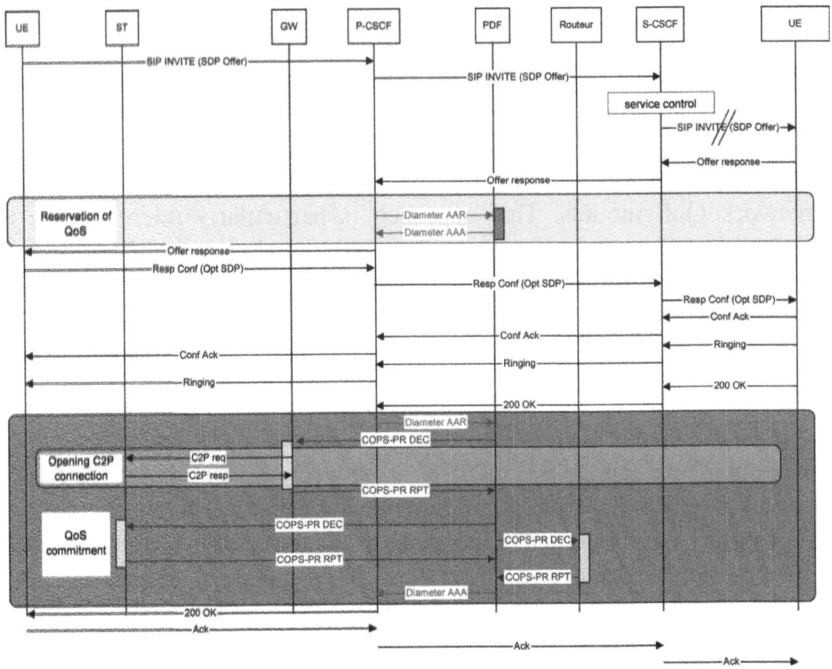

Figure 4.21. *General implementation of QoS with C2P at the level of the NCC for transparent integration*

The establishment (or modification) of the C2P connection is triggered by the NCC upon receiving the COPS-PR decision message. This method avoids adding additional signaling between, for example, the P-CSCF and the NCC or the PDF and the NCC. However, this procedure departs from the procedures defined in the GPRS/UMTS framework where the connection establishment is carried out by the terminal node (section 4.3.2), which triggers in the PEP (GGSN) a COPS request toward the PDF.

In the same way, it is possible to develop a solution where QoS implementation is based on the establishment of a C2P connection by the ST (Figure 4.22). This approach offers an architecture which is independent from the topology of the system in return for a reduced optimization in the session opening delay in the transparent case (double hop). In the IMS context, the ST side of the PEP is responsible for triggering the establishment of the C2P connection.

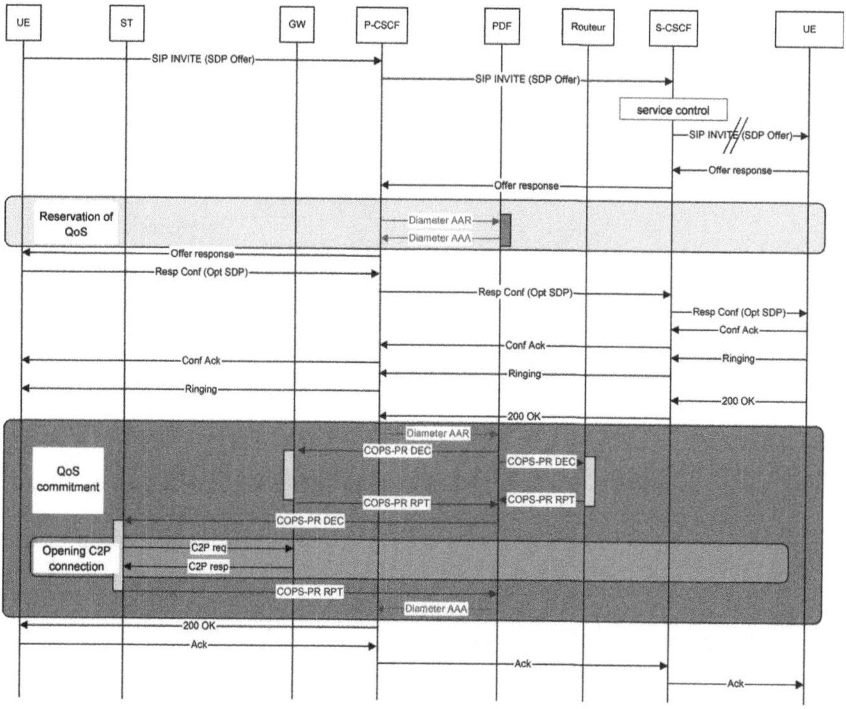

Figure 4.22. *General implementation of the QoS with C2P at the level of the ST for transparent integration*

Of course, the establishment delay is longer here given that the establishment is triggered one satellite hop later than in the previous case.

4.5.2. *Star integration scenario*

4.5.2.1. *Architecture*

In the case of the integrated star scenario, the P-CSCF is located at the side of the ST. As previously indicated, the PDF is centralized and placed at the side of the GW. The position of the PEFs remains unchanged.

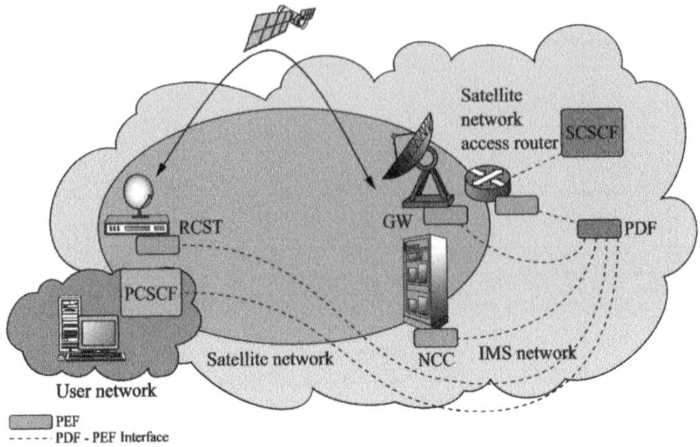

Figure 4.23. *IMS satellite architecture in scenario 2*

4.5.2.2. *Signaling*

Figure 4.24 describes the signaling defined to implement the QoS during the establishment of an IMS session.

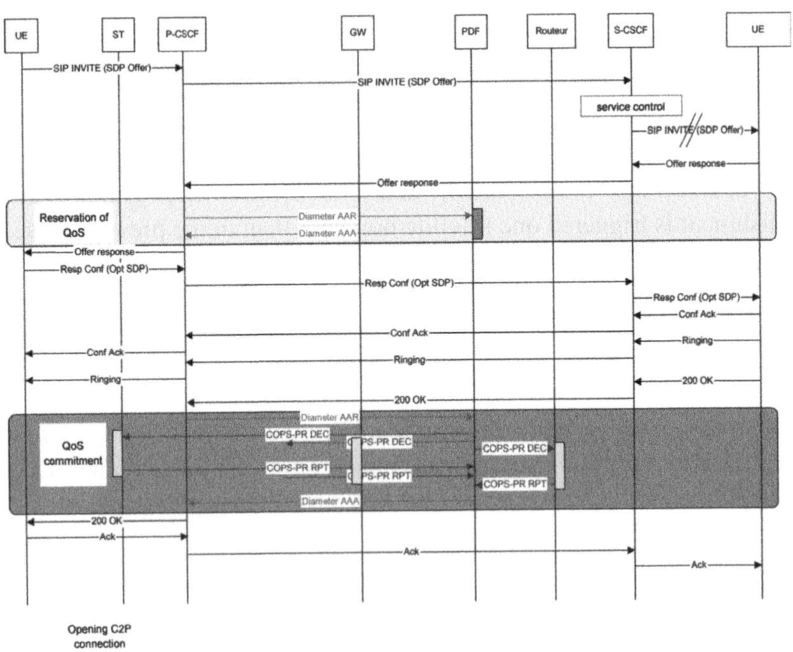

Figure 4.24. *General implementation of QoS for the star integration*

It can be noted that the impact in terms of delay is greater in this case than the last since the PDF is not colocated with the P-CSCF. Two satellite hops are necessary for the QoS reservation, and then four hops are necessary for the actual implementation of the QoS, making four hops more than in case 1.

4.5.3. *Mesh integration scenario*

4.5.3.1. *Architecture*

In the final scenario, the P-CSCFs are placed in each of the user subnetworks served by the ST. One of the STs ensures the connectivity with the home network. This contains the S-CSCF in the case of a separate IMS network, or has a border router which is able to reach the S-CSCF of the IMS network. The PEFs are logically placed in the STs and NCC. It is also possible to place a PEF in the border router to protect the subnetwork.

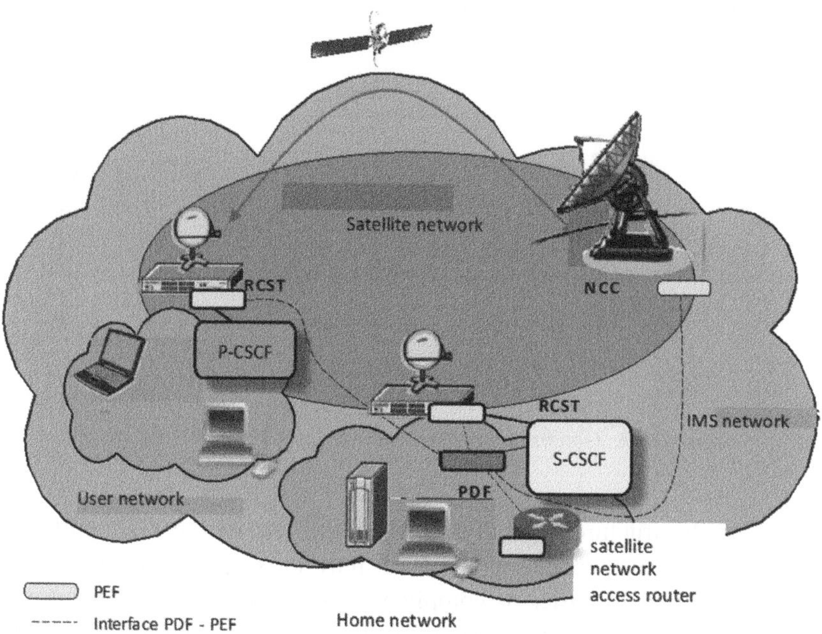

Figure 4.25. *IMS satellite architecture in scenario 3*

4.5.3.2. *Signaling*

In this case, the difference with the previous scenario is the need to configure the access resources (especially, demand assignment multiple access (DAMA)) and thus to communicate directly with the NCC. There is no additional impact in terms of delay if border connectivity is properly provided (that is to say, if the system is not transparent meshed). C2P is not considered here, nor is it chosen as a means of managing QoS at the satellite segment.

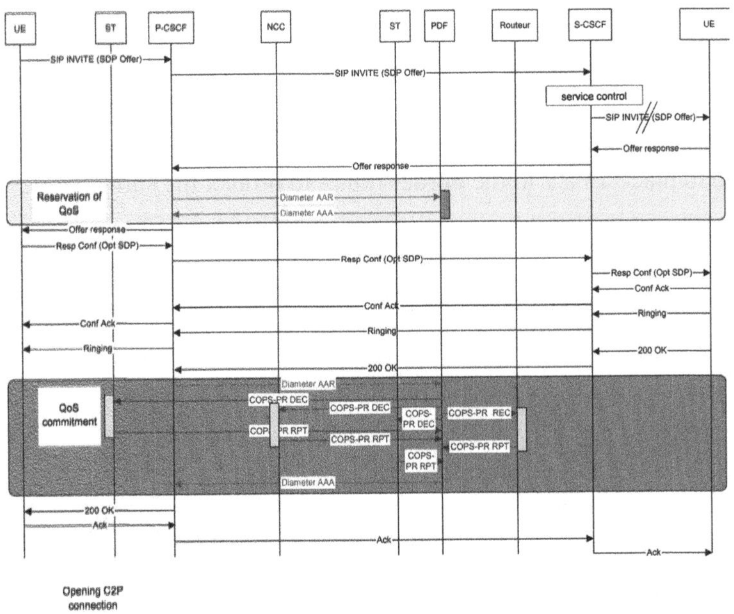

Figure 4.26. *General implementation of QoS for mesh integration*

The mesh structure logically leads to the use of the C2P protocol to establish the access connections, which may have QoS characteristics. As previously stated, in order to comply with the IMS QoS architecture, this case can only be considered if the user terminal is also the ST. If this is not the case, it would be necessary for the user terminal to be able to trigger the establishment of an access connection by specifying the QoS and authentication parameters necessary (e.g. by using SOAP). Unlike the transparent case (section 4.5.1), where a connection is opened by the NCC, it is simpler in this case (topologically speaking) to establish the connection via the terminal node.

Figure 4.27 shows a meshed IMS communication between a user in a visited network and a user in the home network. The establishment procedure is thus similar to that encountered in the case of GPRS/UMTS networks. First, the SIP session initiation phase occurs, and then the reception by the source ST of the OK SIP triggers the establishment (or modification) phase of an access connection. The reception of the C2P-REQ message in the destination ST triggers the QoS implementation request within the access network.

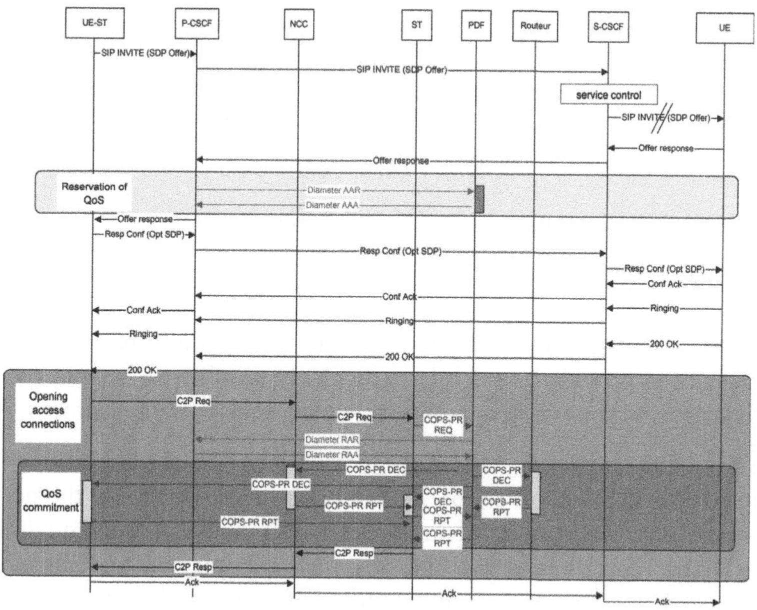

Figure 4.27. *General implementation of QoS for meshed integration with C2P*

It can be noted here that the triggering of the QoS commitment phase can also be considered:

– at the NCC level upon the reception of the C2P-Resp, which ensures that the access connection is not refused after the implementation of the COPS policy. In this case, the establishment delay is increased by one extra satellite hop;

– at the level of the ST source upon the reception of the C2P-Resp, which enables the complete insurance of coherence between the access and network levels. However, the latency delay before starting application streams would also be increased by one satellite hop compared to the initial case.

4.6. SATSIX project

As part of the European research and development funding, the information society technologies (IST) SATSIX project aimed to show the feasibility of this type of IPv6-compatible satellite communication system. End-to-end QoS architecture [RAM 07] based on the high layers (session and network layers), low layers (access and physical layers) as well as cross-layer mechanisms have been defined. Its principle contribution is to support a dynamic QoS, which is QoS management whose parameters are updated on the go according to application requirements.

SATSIX QoS supports two QoS management models:

1) A model known as "QoS enabled", whereby terminal nodes can initiate their application session without the network necessarily guaranteeing a given QoS. In this case, flexible (without restrictions) QoS management is provided, which is particularly suitable for services without significant restrictions (in terms of jitter, delay, transfer rate or drops).

2) A model known as "assured QoS", where the session can only be accepted if the application's QoS demands can be respected by the network. In this case, it is necessary to guarantee the resources and ensure admission control. Extra renegotiation mechanisms may be implemented. This model is clearly suited to interactive applications with high restriction levels (VoIP, video-conference, etc.)

Generally speaking, for high layers, this architecture is based on one or more SIP proxies responsible for collecting session information for dynamically configuring the QoS provided by the satellite segment. Therefore, this information modifies stream-by-stream the parameters of the IP scheduler in the terminal and the configuration of the allocation of the terminal (e.g. by modifying the (DVB-RCS) continuous rate assignment (CRA) or the (DVB-RCS) rate-based dynamic assignment (RBDC max) of the terminal).

This architecture is supplemented by an additional "QoS agent" and a "QoS server" responsible for applying the QoS instructions expressed by the network user or administrator for applications which do not have session signaling or inherent QoS.

At the access level, this architecture is based on level 2 connections in order to transport the QoS parameters. These connections are created, modified and deleted using the C2P protocol. The proposed solution supports both a mesh and a star topology in the same way. This also allows only one SIP proxy to be involved in the terminal.

The "Access Resource Controller" has an interface between the SIP proxy and the DAMA of the NCC in order to make dynamic modifications of the allocation parameters. This component is also designed to offer a generic entry point to other proxies.

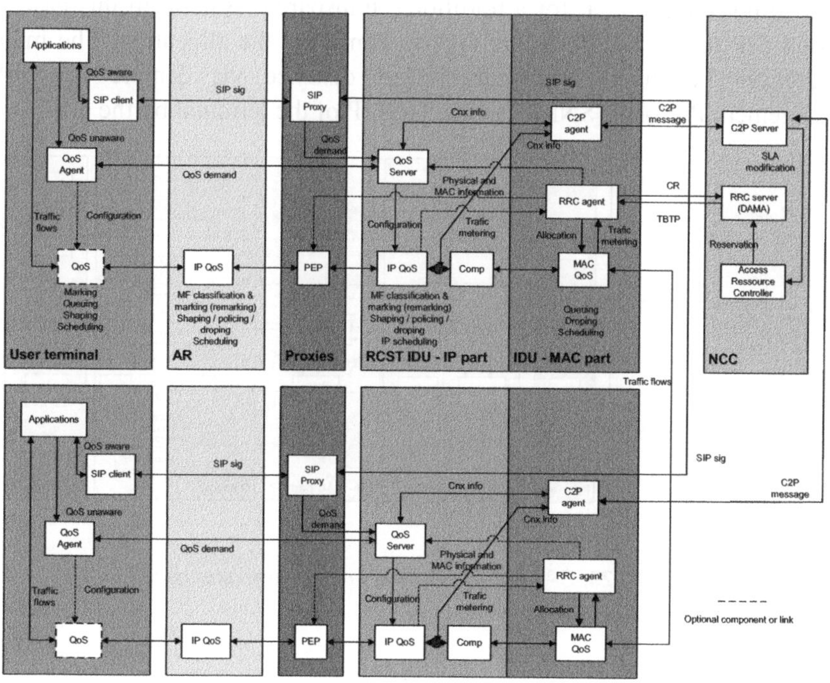

Figure 4.28. *Access-oriented SATSIX architecture (mesh case)*

Moreover, different optimizations have been proposed, notably via the definition of a hierarchical dual token bucket (HDLB) scheduler, ensuring QoS at the network level. HDLB comes from hierarchical token bucket (HTB), from which it takes the hierarchical representation of service classes by adding to each branch a management based on the dual leaky buckets. This new scheduling technique enables a better control of bursts and a better sharing of traffic between classes. Finally, a feedback loop was defined between the IP and MAC layers in order to control the transfer rate of the IP scheduler according to the size of the MAC queues (and thus according to their congestion). This technique improves the coherence of the scheduling decisions taken within the two layers: in an advanced way in layer 3 and in a simple way in layer 2. It also addresses the variable link issue (modulations, coding and symbol timing can be dynamically modified).

This architecture has also been adjusted (in an IP-oriented approach) in order to not require C2P for a traditional transparent system. In this case, a second SIP proxy is used in the GW to configure the allocation. The main differences lie in the configuration at distance via Simple Network Management Protocol (SNMP) of the QoS IP of the terminal by the GW.

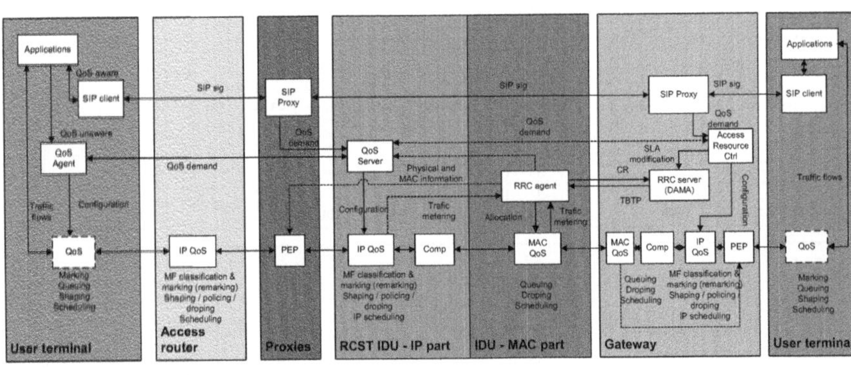

Figure 4.29. *SATSIX IP-oriented architecture (star case)*

4.7. Conclusion

The use of the QoS architecture defined in the IMS framework is perfectly applicable in the satellite case. The impact in terms of delay of implementation of QoS is marginal compared to the delay in establishing the

IMS session. The architecture known as "transparent integration" is the most relevant in terms of efficiency for issues related to QoS. Moreover, the lack of equivalence with C2P contexts simplifies the implementation and the number of procedures to manage the QoS.

This architecture extends the concepts which had been defined and implemented as part of the European SATSIX project. Therefore, the IP-oriented approach (not requiring C2P in the case of a traditional transparent system) is mainly based on SIP to allow a vertical optimization of QoS within the satellite segment (see Figure 4.29).

With regard to this architecture, the function ensured by the SIP proxy colocated in the GW is taken on by the P-CSCF. The active resource controller (ARC) mainly plays the role of PDF, although it also has an "enforcement"-type function via the configuration of the allocation carried out by the DAMA. The QoS server is a QoS control point (PEF) which would be activated in the new architecture for COPS instead of the second SIP proxy located in the ST. Compared to the architecture partially implemented as part of the SATSIX experiments, where the configuration of IP QoS is carried out at a distance via SNMP without SIP proxy in the terminal, the SNMP protocol is replaced by COPS, which is more appropriate for QoS management.

The architecture proposed is relatively easily integrated into the broadband satellite multimedia (ETSI) (BSM) quality of service (QoS) architecture (as shown in Figure 4.30), mainly due to the opening of this architecture. Therefore, the service control function (SCF) is carried out by P-CSCF, S-CSCF, (IMS) home subscriber server (HSS) and AF elements, and the decision functions (PDP) and QoS implementation (PEP) are already present. Nonetheless, it may be noted that the IMS PEFs are responsible for the configuration of gates for the authorized services.

As well as the generalization of the previous two architectures, the IMS QoS allows a horizontal integration of heterogeneous segments, which is hardly addressed at all in the SATSIX and BSM architectures. Finally, it integrates functions, above all of authentication and access control of services, which are either absent (SATSIX) or only partially addressed (service control in the BSM architecture).

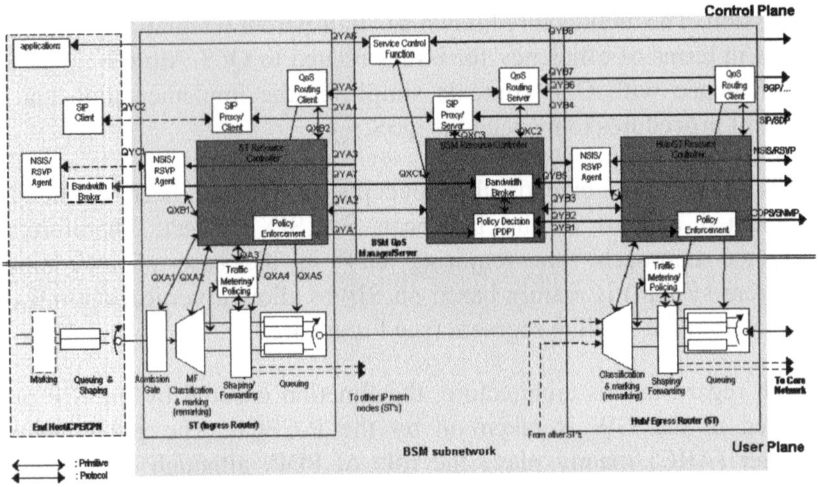

Figure 4.30. *BSM QoS architecture*

5

Inter-system Mobility

5.1. Introduction

For several years, digital video broadcasting – return channel via satellite (DVB-S/RCS) satellite systems have made significant progress in the field of broadcasting and coding techniques, as well as in the implementation of a satellite return channel, providing users with bidirectional access. This has allowed users to become competitive as access networks for isolated or low-density regions and for company networks no longer require heavy and costly infrastructure. However, it has become important for this type of network to be able to offer its users the same services as those provided by terrestrial networks. Indeed, the appearance of numerous multimedia applications and the ever-increasing mobility of users have introduced new obstacles to be overcome.

Many studies have been undertaken to tackle this growing demand. However, the solutions offered for terrestrial networks do not always prove themselves to be suitable for the specific constraints of satellite systems which suffer from long propagation delays, limited bandwidth and more frequent transmission errors. The implementation of a suitable architecture thus becomes crucial to the development of these networks, especially in terms of quality of service (QoS) since the time requirements imposed by the multimedia applications combined with the difficulties introduced by the management of mobility must be taken into account.

First, this chapter will examine the issue of mobility in satellite systems by explaining the main types of mobility which exist. Section 5.3 provides a state-of-the-art of the various protocols developed to manage mobility in terrestrial networks. Section 5.4 offers two solutions at different levels providing a mobility of service in the satellite networks which is compatible with terrestrial solutions. The performance of these solutions will be analyzed to ascertain the differences between them and the service which they can deliver. Finally, section 5.5, which takes a wider view, offers a solution for the provision of a mobile service with guaranteed QoS, compatible with the IP multimedia subsystem (IMS) approach of third generation partnership project (3GPP).

5.2. The taxonomy of mobility

In order to avoid confusion, the first part of this chapter presents various concepts incorporated by the term "mobility". We will attempt to differentiate the main types of mobility as they have been defined over recent years.

5.2.1. *Personal mobility*

The notion of personal mobility is used to describe the possibility of reaching a user with the same logical identifier wherever he/she is, whatever the communication device or terminal (personal computer, laptop, mobile telephone, etc.) he/she is using, regardless of the access technology (global system for mobile communications (GSM), WiFi, Ethernet, etc.). Therefore, the same identifier can be associated with various terminals and several identifiers can be associated with the same terminal as shown in Figure 5.1.

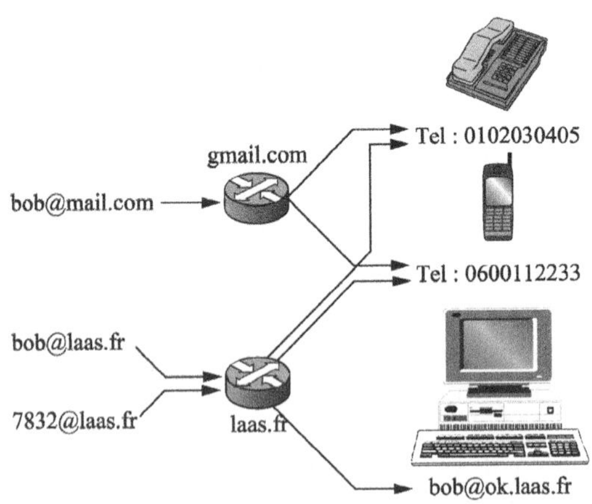

Figure 5.1. *Example of personal mobility*

In this example, Bob can be reached from the outside on any of his terminals via different identifiers (e.g. email address) in a transparent way for the person attempting to make contact. He may also wish his personal identifier (bob@laas.fr or 7832@laas.fr) to be redirected to another telephone, and also for his private identifier to only be redirected to his home telephone and smartphone.

5.2.2. *Session mobility*

Session mobility must allow a user to keep his/her sessions active while changing terminal. Therefore, a user with voice over IP (VoIP) communication on his/her smartphone may, upon arriving at the office, decide to continue the conversation on his/her PC without there being a break in the communication.

5.2.3. *Mobility of service*

The mobility of service must allow the user to access the same services to which he/she is subscribed with his/her access provider wherever he/she is (possibly via another access provider) and whatever type of terminal or technology he/she is using. Therefore, using the example of VoIP, the user, wherever he/she may be, will have access to his/her list of contacts, call blocking, media preferences and more generally to all the options contained in the service subscription.

5.2.4. *Mobility of the terminal*

The mobility of the terminal must allow a user to keep sessions active and to remain contactable from the outside while changing Internet Protocol (IP) network or subnetwork (and therefore access point (AP) or connection point). In this case, the user does not change terminal during the session. In [MAN 04], and then in [KEM 07], the internet engineering task force (IETF) defined three distinct mobility categories for mobile nodes (MNs):

1) *Link mobility or level 2 mobility* which defines mobility between two wireless APs of the same access network. Typically, this kind of mobility only involves level 2 mechanisms or means at least that no reconfiguration of the network or IP subnetwork is necessary. However, signaling at the

network level or higher may be required for the exchange of messages between APs. This type of mobility is also known as horizontal mobility or intratechnology.

2) *Micromobility* or *local mobility* which is mobility within the same access network but which involves network or IP subnetwork reconfiguration mechanisms. However, although terms such as "micro" and "local" are used, this does not mean that the geographic area covered by this access network in itself cannot be large. It can be considered as localized management of mobility as signaling, to maintains the IP connectivity, is restricted to the access network.

3) *Macromobility* or *global mobility* which is mobility between different access networks, involving an IP reconfiguration, without taking account of the type of technology. The management of mobility cannot be done locally and can involve significant changes at the level of end-to-end packet routing.

These types of mobility may be implemented in the case of inter-system mobility.

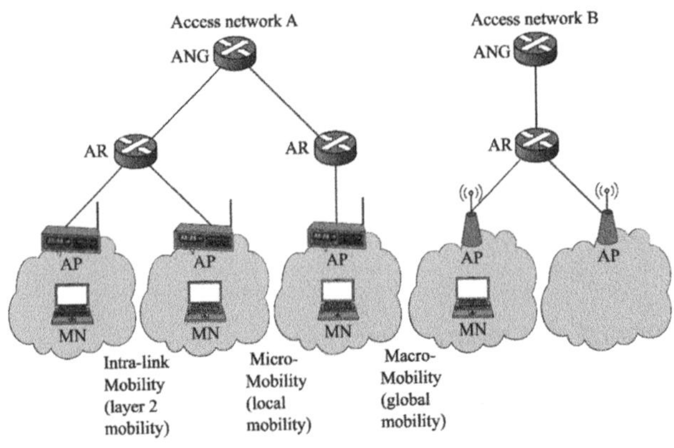

Figure 5.2. *IETF mobility terminology*

Finally, the mobility of the terminal can be divided according to its "granularity", by distinguishing three categories:

– *discrete* or "*nomadic*" mobility is applied to a user who is moving and therefore who changes connection point without communication in progress;

– *continuous* mobility refers to a user who is moving while communication is in progress. He/she remains contactable while moving but his/her communication may be interrupted due to the various stages of connection to the new AP, the allocation of a new IP address, etc.;

– *uninterrupted* or "*seamless*" mobility refers to a user who is moving while the communication in progress does not suffer the slightest interruption.

5.2.5. *Network mobility*

Network mobility refers to a set of networks or subnetworks connected to the Internet via the intermediary of one or more mobile routers changing their Internet connection point. This type of mobility can be used, for instance, when a network inside a vehicle (train, boat, car, bus, plane, etc.) wishes to access IP services during travel. These in-built networks onboard vehicles may different in nature: for example, a sensors network deployed on a boat which exchanges data necessary for navigation, or an access network on a train allowing passengers to connect during travel.

5.2.6. *Clarification for mobility terminology*

As demonstrated here, the term "mobility" can have a large number of different meanings, making it a complex issue. This chapter will focus on terminal mobility, which is currently the subject of the majority of the research being carried out and which poses the most significant problems. Nonetheless, this does not prevent us, when the opportunity arises, to explain how such a solution may also enable the management of another type of mobility. However, if no type is specified, the term "mobility" refers in this chapter to *terminal mobility*. Moreover, we prefer to use the distinction {level 2 mobility/micromobility or local mobility/macromobility or global mobility} rather than using the horizontal/vertical distinction.

5.3. Protocols for mobility management

The mobility of the user, or his/her terminal, can be managed in the communication system at different levels, which entails extra complexity in the layers in question. Therefore, mobility can be managed at the lowest layer, via the adaptation of physical and link layers, via the network layer by implementing a mobility protocol in layer 3 or even in the higher layers, such as the transport and session layers. This section examines the different alternatives, particularly focusing on the protocols of the network and session layer, whose solutions are described in the subsequent sections of this chapter.

5.3.1. *Extension of DVB-RCS for mobility*

DVB-RCS2 approved by the DVB consortium in 2011 was extended in 2012 to include mobility management [ETS 12c], initially known under the name DVB-RCS2+M. The management of mobile and nomadic terminals is included in the standard as well as mechanisms enabling hand-overs between spots. The mobility of spots is ensured by the lower layers of the system, since the mobility of the gateway (GW) or satellite must be processed at higher protocol levels. Forward error correction (FEC) mechanisms are defined to limit masking effects of satellite link in mobility.

A distributed mechanism for managing hand-over is set out in the standard, offering flexibility and simplicity. Hence, the satellite terminals (STs) initiate the hand-over, but the final trigger is made by the network control center (NCC) responsible for the terminal which made the request. Nevertheless, a centralized mechanism, under the full responsibility of the NCC, is also set out.

In the first case, the hand-over procedure can be broken down into three phases: detection/signaling, decision and execution. The ST is responsible for the detection and signaling mechanisms, while the NCC needs to manage the final two phases. After detecting a requirement for a hand-over, the return channel satellite terminal (RCST) must signal its request to the NCC via "Mobility_Control_Messages", containing the request and the priority spots requested. The request may address the forward and return links, while two requests will be emitted in the latter case. The request should be emitted at least 10 s before the break of the connection. In the event of multiple spot requests, the NCC will offer a special hand-over solution.

In the second case, the NCC with the information it has must be able to detect the need for a change, for example with the information obtained at the connection of a terminal and the different tables updated by the system. The NCC sends a hand-over request to the terminal, as well as all the information necessary for the operation in the spot chosen by the terminal information message – unicast (TIM-U). This message contains a mobility descriptor, a descriptor of the forward connection (identity of the new spot and physical characteristics of the connection), a descriptor of the return channel (sequence number of the superframe and physical information on the return channel), a control descriptor indicating the information generally transported to a (DVB-RCS) terminal time burst time plan (TBTP) frame as well as a logon descriptor (group identifier (ID) and login ID). Upon reception, the terminal must be able to synchronize itself with the new spot, to stop sending control bursts in the previous spot and to start a logon in the new spot (the latter will be ignored by the NCC).

The GW hand-overs should respect similar mechanisms which are not detailed in the standard.

The spot mobility managed at this level is transparent for higher layers and must have a limited impact on the system. However, if the system does not have this management level, it is possible to manage the spot mobility in the higher layers as proposed in the following section.

5.3.2. *Management by the network layer: mobile IP*

The most common implementation on the network layer is the IP. It is mainly responsible for routing the packets from the source to the destination. Currently, this protocol is deployed under the name IPv4 and enables the use of slightly more than 4 billion different addresses. However, the growing success of the Internet could mean that in the years to come this number may not be sufficient despite the use of solutions such as network address translation (NAT) and division into address classes. It is largely for this reason of address shortage that a new protocol was developed: IPv6.

Since mobility is not included in IPv4, it seems unlikely that the modifications necessary for its implementation will be carried out by this protocol, or in any case, not on a large scale. For this reason, in the next section, we will focus on mobility protocols which work with IPv6.

5.3.2.1. *Mobile IPv6*

In 1994, benefitting from the emergence of the new IPv6 protocol and improving the mechanisms defined in Mobile IPv4 [PER 02], three researchers submitted to IETF a proposal for a mobility protocol on IPv6 called Mobile IPv6, which describes a way of managing the mobility of IPv6 terminals. However, disagreements regarding the security of Mobile IPv6, as well as the different optimizations possible, made standardization a long and laborious process, and it was only in June 2004 that a request for comments (RFC) 3775 was finally published.

5.3.2.1.1. Terminology

Mobile IPv6 describes a mechanism involving three entities: the mobile IPv6 terminal (MN) which can move from one network to another and therefore change its connection point to a network or subnetwork to another; the home agent (HA) responsible for redirecting the packets destined for the MN when it is in a visited network; and the correspondent node (CN) which communicates with the MN. Therefore, three types of network can be distinguished:

– the home network, which is the home network of the MN;

– the correspondent network, which is the network of the CN (which may be the same of the home network);

– the visited network which is a network (other than the home network) the MN moves to.

To enable the MN to always remain addressable, a permanent address, known as the home address (HoA), is assigned to it. Moreover, when it is in a visited site, it is allocated a temporary address, care-of address (CoA).

5.3.2.1.2. Basic principle

When the MN is in the home network (see Figure 5.3(a)), routing is carried out in the standard way, based on the routing tables, since in its home network, the MN acts as a "fixed" IPv6 terminal. When the MN moves in a visited network (see Figure 5.3(b)), it takes on a CoA whose prefix will obviously be that of the visited network. To do this, upon its arrival in the visited network, it either directly receives an unsolicited *router advertisement* (RA), or it sends a *router solicitation* to force the RA to be sent. Due to this message, it ascertains the prefix of the network and can therefore build itself an address due to the IPv6 autoconfiguration mechanism IPv6 [THO 98] (combination of the prefix and the medium access control (MAC) address). Once the duplicate address detection (DAD) mechanisms [THO 98] are implemented, the MN registers its CoA with its HA by sending it a binding update (BU) which consists of both its home address and temporary address. The MN then waits for the response from its HA via the binding acknowledgment (BACK). The HA then plays the role of proxy (see Figure 5.3(b)): the packets sent by the CN (which is transparent for the mobility of MN) to the MN are intercepted by the HA which encapsulates them and “channels” them to the destination of the CoA of the MN, while the MN transmits its packets to the CN via the HA.

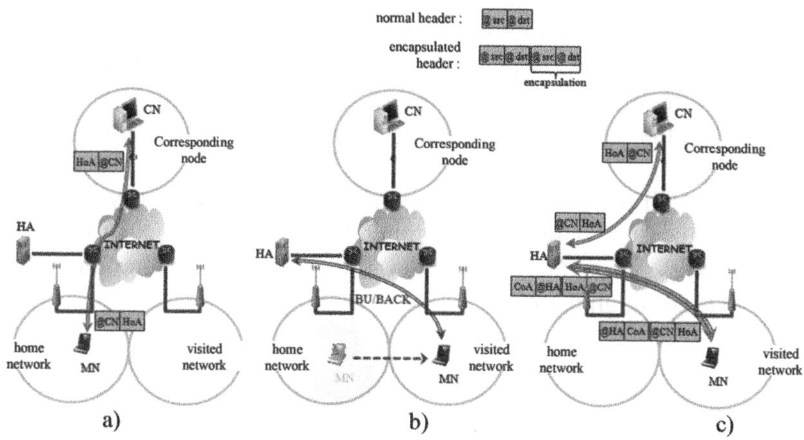

Figure 5.3. *Implementation of the bidirectional tunnel in the Mobile IPv6 a) direct communication, b) binding update with the HA and c) communication in bidirectional tunnel mode*

5.3.2.1.3. Routing optimization and procedure for the return routability test (RRT)

One of the main improvements brought by Mobile IPv6 lies in the fact that the MN can inform its correspondent of its temporary address by exchanging BU/BACK messages with it directly. Indeed, the systematic routing by HA of the mobile remains particularly inefficient at network level, although it is simple to implement and very safe since communication between the HA and the mobile is secured by IPsec. For instance, if the mobile moves far from its home network and communicates with a server close to it, it is more efficient to communicate directly than to go via the HA. This saves on Internet resources, especially for the home network. Furthermore, if a HA needs to redirect the packets of a large number of MNs, it may not be able to handle the load.

When the MN receives a packet encapsulated by the HA initially coming from the CN, it can decide to signal to the CN its CoA with the exchange of BU/BACK messages (see Figure 5.4(b)), in the same way as with HA. This allows the CN to communicate directly with the MN due to the use of two IPv6 options, the routing header, type 2 and the destination option header, which are added to each IPv6 packet to indicate the HoA of the MN (see Figure 5.4(c)). The packets are then directly routed from the CN to the MN (and vice versa), but when the MN receives the packet addressed to its CoA, it collects from the routing header its HoA which is used as a final destination address of the packet. This is what allows the Mobile IPv6 to be transparent for applications in "optimized" mode. However, Mobile IPv6 as well as the routing optimization must be implemented at the CN level.

Indeed, not all the potential IPv6 correspondents necessarily support the routing optimization. If no route optimization is possible, a correspondent responds that is does not understand the BU and the communications continue to go through the HA.

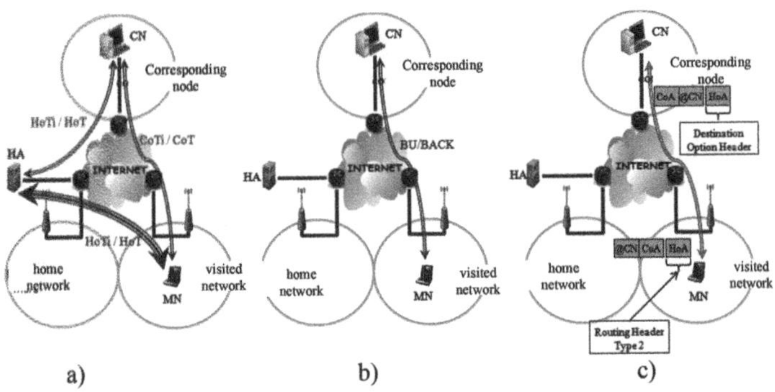

Figure 5.4. *Routing optimization procedure in Mobile IPv6: a) procedure for the return routability test; b) binding update with the CN and c) direct communication with specific routing options*

Another problem emerged during the design of the routing optimization phase of the Mobile IPv6 protocol: the BU mechanism raises significant security problems. Indeed, it is easy to protect the signaling exchanges between the MN and HA because of the administrative relationship which, for example, enables the use of a shared secret. It is much more complicated with regard to the correspondents and yet the security of binding updates is crucial. Without protection, it would be possible to divert communications from a mobile by redirecting traffic in order to spy on it or to carry out a denial-of-service attack. This is why a procedure known as RRT is specified in order to protect signaling between the MN and its correspondents.

5.3.2.1.4. Mobile IPv6: conclusion

Despite the improvements made to the routing optimization mechanisms, Mobile IPv6 still suffers from a latency leading to the dropping of important packets which can be detrimental, especially for "real-time" applications. This latency is mainly due to three factors:

– level 2 hand-over delay (detection of network change and association of new AP);

– the configuration of a new CoA (obtaining an address and DAD);

– BU/BACK exchanges and messages necessary for the RRT procedure with the HA and CN.

A number of solutions are presented in the following sections in order to overcome these delay-related problems.

Mobile IPv6 also suffers from another problem due to the central role which HA plays in mobility mechanisms. Indeed, HA malfunctioning (crash, denial-of-service attack, no route available to HA, etc.) would lead to a complete interruption of MN communications if the routing optimization was not carried out, and in any case, would make the conservation of communications in the event of a network change impossible. Various reliability mechanisms have thus been proposed by IETF based on the principle of HA redundancy. However, the disadvantage of this is that it makes Mobile IPv6 even more complex and heavy in terms of necessary infrastructure and therefore has yet to be standardized.

Finally, congestion problems from this central role played by the HA can arise when a large number of MNs are associated with the same HA.

5.3.2.2. *FMIPv6: Mobile IPv6 fast handovers*

FMIPv6 [KOO 09] is certainly one of the most promising improvements produced for IPv6 mobility. The objective of this protocol is to reduce the hand-over delay by improving on one hand the movement detection time of the MN and on the other hand the registration time of the new CoA. FMIPv6, therefore, defines the new mechanisms, independent of the level 2 technology to enable the MN:

– to configure an IPv6 address for the next network before moving effectively;

– to send packets as soon as a new link is detected (due to the preconfiguration of an IPv6 address);

– to receive packets as soon as its new access router (AR) detects its attachment (via tunnel and buffering mechanisms).

These mechanisms are fully compatible with Mobile IPv6 (or, as the standard specifies, with other protocols managing IPv6 mobility, but in this section, we will assume that Mobile IPv6 is being used).

In the best-case scenario, the overall interruption time can be reduced to the interruption time due to level 2 rebinding.

New entities emerge in order to do this: the previous access router (PAR) and the new access router (NAR) which correspond to the ARs before and after the hand-over procedure, respectively. Moreover, two temporary addresses are defined: the previous CoA (PCoA) and the new CoA (NCoA) corresponding to temporary addresses obtained from the PAR and NAR, respectively. The reference architecture for FMIPv6 is given in Figure 5.5.

When the MN is still connected to the PAR, it can obtain information on possible future APs to which it can connect due to level 2 mechanisms. The principle of FMIPv6 is, therefore, to allow the MN to request its PAR for information regarding the ARs which correspond to one or more APs discovered by sending to the PAR an router solicitation for proxy advertisement (RtSolPr) internet control message protocol (ICMP) message whereby it indicates the identifiers (MAC address) of the APs discovered. In response to this message, the PAR sends another ICMP message, this time a proxy router advertisement (PrRtAdv), in which it indicates the information regarding surrounding ARs corresponding to the APs indicated. This information takes the form of (AP-ID and AR-Info) and the AR-Infos may contain:

– the MAC address of the new router;

– the IP address of the new router;

– the IPv6 prefix given by the new router;

– possibly a NCoA if the PrRtAdv is sent by the PAR without solicitation from the MN. If this option is present, the MN must immediately send a fast binding update (FBU) if it does not want to risk a loss of connectivity.

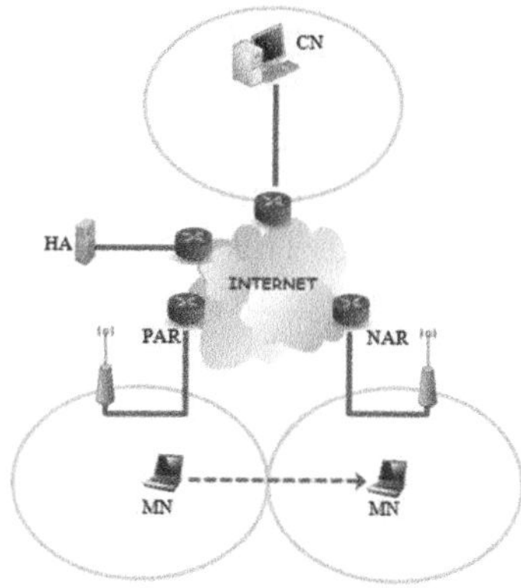

Figure 5.5. *FMIPv6 architecture*

Once the PrRtAdv message has been received by the MN, two operating modes are defined by FMIPv6: predictive and reactive modes.

5.3.2.2.1. The FMIPv6 predictive mode

When possible, as soon as the MNs receive the PrRtAdv (and an NCoA is configured where necessary), they send an FBU to the PAR directly via the corresponding level 2 link. The predictive mode (see Figure 5.6) is used when the MN receives a fast binding acknowledgment (FBACK) from the same connection. If it does not, the MN considers the FBU as lost and sends it another time from the new link, therefore, corresponding to the reactive mode.

When the PAR receives the FBU in which the PCoA of the MN is indicated along with the NCoA as an *alternate CoA option*, it sends to the NAR a *handover initiate* (HI) message containing the MAC address of the MN, its PCoA and the NCoA which the MN wishes to use. The NAR must, therefore, carry out the DAD procedure before responding with a *handover acknowledge* (HACK) message which specifies whether the handover has been accepted or not and, if it is accepted, it also contains the NCoA that the MN should use. The PAR then sends an FBACK to the MN by specifying this NCoA. A tunnel between the PAR and NAR is, therefore, implemented: the PAR intercepts the packets destined for the PCoA of the MN and sends them to the NAR which can either send them on to the NCoA of the MN if it is already attached, or buffer them. The MN, once the FBACK is received, can change the connection and attach itself to the NAR. Once it is attached, the MN should immediately send an *unsolicited neighbor advertisement* (UNA) message to the NAR which can then send its packets. The MN can then also start again to send packets via the reversed tunnel (e.g. MN→NAR→PAR→CN if the communication is made directly between the MN and CN before the movement). This bidirectional tunnel maintained for long enough for the MN to exchange BU/BACKs with its HA and possibly its CNs.

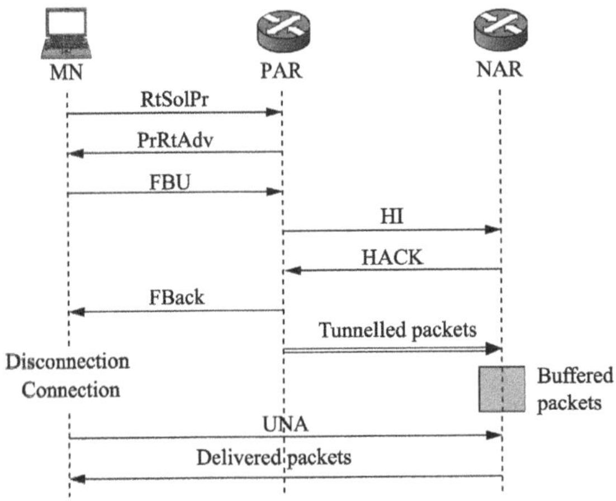

Figure 5.6. *FMIPv6 in predictive mode*

5.3.2.2.2. The FMIPv6 reactive mode

The FMIPv6 operating mode is known as reactive when the MN cannot send the FBU from the link corresponding with its PAR or in the case where the MN sends the FBU from this connection, but receives no FBACK before changing the attachment network. Once it is attached to the NAR, the MN sends a UNA message to the NAR immediately followed by an FBU (even if the MN has already sent one from the previous network, there is no way of knowing whether the PAR has been received or not) to the PAR, and the NAR simply passes the message on. Subsequently, the HI/HACK messages are exchanged in the same way as in the predictive mode. The PAR can thus tunnel the FBACK as well as the packets toward the MN via the NAR until the exchange of specific Mobile IPv6 messages is over. NAR to MN direction, as soon as the MN receives the FBACK, it can start once more emitting packets via the bidirectional tunnel during the exchange of Mobile

IPv6 messages. Figure 5.7 summarizes the messages exchanged by FMIPv6 in reactive mode.

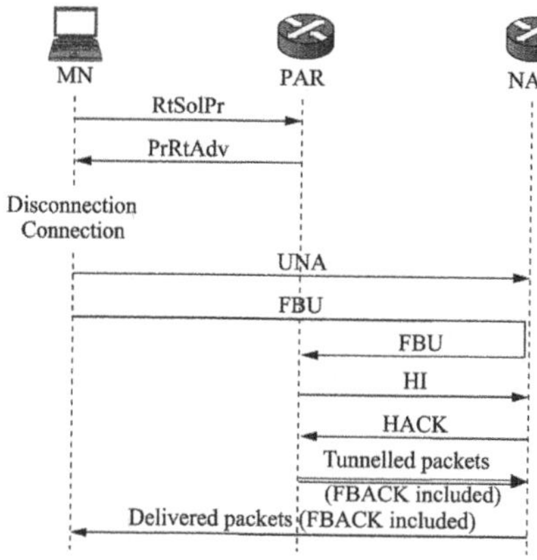

Figure 5.7. *FMIPv6 in reactive mode*

5.3.2.2.3. Hand-over initiated by the network

In some wireless technologies, the hand-over can be initiated by the network and not by the mobile. In this case, the PAR sends an unsolicited PrRtAdv message containing the MAC address, IP address and subnetwork prefix of the NAR. The MN must, therefore, configure an NCoA and send an FBU to the PAR. The rest of the operations follows one or the other of the two modes previously described according to whether the MN receives the FBACK from the link corresponding to the PAR or not.

5.3.2.3. *HMIPv6: hierarchical Mobile IPv6*

The Mobile IPv6 mechanisms appear to be relatively inefficient when the mobile changes its Internet connection point within the same domain, especially when the distance traveled by the MN is short in comparison to the MN/HA and MN/CN distance. Indeed, each movement requires at least one interaction between the host network and the HA (BU/BACK). Moreover, if the MN is in communication with several CNs and wishes to re-establish direct communication with them, it must exchange a high number of messages which, in the framework of wireless technologies, for example, can be detrimental in terms of load in the system.

HMIPv6 [SOL 08] was then proposed in order to enable better management of these movements within a domain, based on a hierarchical mobility management model. To do this, HMIPv6 proposes the use of a new entity: the mobility anchor point (MAP). This is a router located in the domain made up of several host networks and used as a local HA by the MN. If the MN moves within the MAP domain (local domain in which mobility is managed by an MAP), the signaling necessary for IPv6 mobility is limited to this domain and thus becomes transparent for the HA and CN(s). HMIPv6 also defines two new temporary addresses: the *regional CoA* (RCoA) allocated by the MAP to the MN and the *on-link CoA* (LCoA) allocated by the current AR to which the MN is attached.

It is important to highlight that HMIPv6, just like FMIPv6, is fully compatible with Mobile IPv6. The MN can, therefore, choose whether or not to use HMIPv6 to manage its mobility (e.g. if it is in a home network or a network close to the home network, it can opt to use its HA rather than an MAP but the objective of this section is not to specify when this choice must be made). However, HMIPv6 may be used independently from Mobile IPv6 (without HA), with successive MAPs playing the role of HA. The HMIPv6 reference architecture is shown in Figure 5.8.

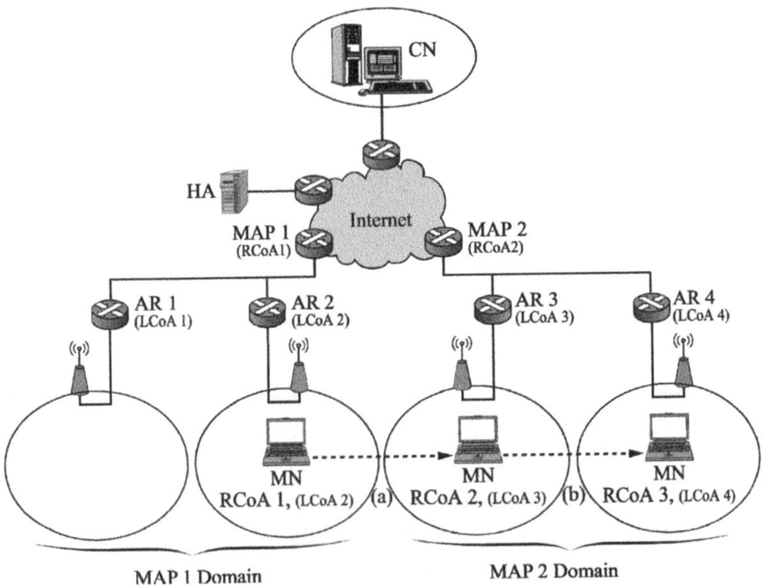

Figure 5.8. *HMIPv6 architecture*

5.3.2.3.1. Mobility with MAP domain change

Although the use of HMIPv6 is mainly interesting for mobility within an MAP domain, it also enables its management during domain changes. Therefore, when MN arrives in the new MAP2 domain (movement (a) in Figure 5.8 and 5.9), it must configure its new RCoAs and LCoAs. To do this, when it receives an RA from AR3, it locates the MAP option (this option must be manually configured by the network administrator at the level of the ARs and the messages it receives) which contains one or more MAP IP addresses. The MN then chooses the MAP which has the highest preference value. In our example, it obtains the MAP2 IP address which allows it to configure its (HMIP) regional care-of-address (RCoA2) as described in [DEV 07]. It also constructs its LCoA3 by traditional IPv6 autoconfiguration mechanisms.

The MN then informs the MAP2 of its new association between the RCoA2 and the LCoA3 with a *local binding update* (LBU). The MAP2 then responds with a BACK containing a *routing header, type 2* which includes the RCoA2. After being registered by its MAP, the MN must inform its HA and possibly its CN by exchanging BU/BACKs specifying the association between its RCoA2 and HoA, in the same way as for Mobile IPv6. Once these operations have been completed, the MAP2 plays the role of proxy by intercepting the packets destined for the RCoA2 and encapsulating them for the destination of the LCoA3 of the MN in the same way as the HA does with the HoA and CoA.

The MN is finally able to communicate directly with its CN (without going via the MAP, if, for example, the CN is located in the same subnetwork) by sending it a BU message specifying the LCoA3 – HoA association, but this possibility is not shown in Figure 5.9.

5.3.2.3.2. Mobility within the MAP domain

The advantage of using HMIPv6 lies in the movements within the same MAP domain. Indeed, in this case (movement (b) in Figure 5.8 and 5.9), the MN just needs to send an LBU to the MAP2 to inform it that it now has the local LCoA4 address. Therefore, it is not necessary to send a message to the HA and CN. HMIPv6 may, therefore, typically be used as a localized management solution for mobility in the same access network. However, this involves IPv6/IPv6 encapsulation between the MAP and MN which may be detrimental to the wireless link used by the MN.

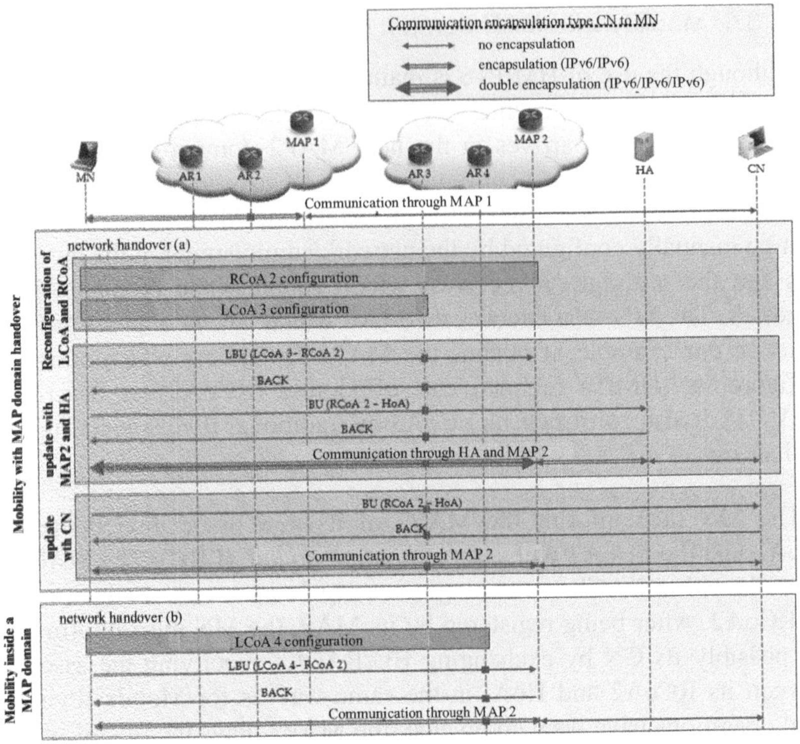

Figure 5.9. *Mobility management by HMIPv6*

5.3.2.3.3. Combination of HMIPv6 and FMIPv6

Although it is efficient in intra-MAP domain movement, HMIPv6 is even longer than Mobile IPv6 in its management of this type of movement. Indeed, it adds an extra LBU/BACK exchange between the MN and MAP.

To reduce the impact of this type of movement, HMIPv6 advises sending an LBU to the previous MAP (MAP1 in our example) by specifying its new LCoA. This can then send packets to the new location of the MN, if authorized (e.g. within the same administrative domain) but this stage is not specified, and concurs with the FMIPv6 principle.

The idea of combining HMIPv6 with FMIPv6 has, therefore, emerged. This combination, also known under the name of F-HMIPv6, can be carried out in a number of ways as specified by the previous RFC corresponding to HMIPv6 in its appendix. A first solution consists of implementing a tunnel, as in FMIPv6, between the MN, PAR and NAR. However, this would lead to a double pathway between MAP and PAR. To avoid this, a second solution has been proposed of implementing a tunnel between the MN, MAP and NAR.

By applying this final combination to our example in Figure 5.8, in the case of movement (a), the gain in terms of hand-over time is significant for HMIPv6 since the packets are retransmitted from the PAR or the MAP during the level 2 hand-over, while FMIPv6 only gains the double pathway for the MAP-PAR connection. Nonetheless, for movement (b), the two protocols benefit significantly from one another because as well as avoiding the useless double pathway, level 2 latency time is removed by FMIPv6 and the MN does not need to update its associations with the MAP rather than with the HA and CN.

5.3.2.4. *PMIPv6: proxy Mobile IPv6*

Another proposal regarding IPv6 mobility has been made for management to be fully managed by the network. After several years of research, the IETF NetLMM Working Group thus produced a standard under the name proxy Mobile IPv6 [GUN 08]. The requirement for a solution to mobility based solely on the network was initially addressed by Kempf [KEM 07], explaining that this would allow any MN to move between different access networks without needing to implement specific mobility solutions itself.

5.3.2.4.1. Basic principles

PMIPv6 defines the notion of a PMIPv6 domain where mobility is managed by this protocol. An MN located within a PMIPv6 domain is connected to a *local mobility anchor* (LMA) which adopts the role of its HA. Moreover, when it moves, the MN attaches to successive *mobile access gateways* (MAGs) which play the role of ARs responsible for signaling related to its mobility. The PMIPv6 principle is thus to emulate the fact that the MN is always in its home network assigning it a *home network prefix* (HNP), solely assigned to this MN. The MN, therefore, considers the PMIPv6 domain as a single connection. PMIPv6 also specifies that the LMA can assign several HNPs to the same MN (in the same way, they will only be assigned to this MN) but subsequently, we will assume that a single HNP is assigned. The PMIPv6 reference architecture is shown in Figure 5.10.

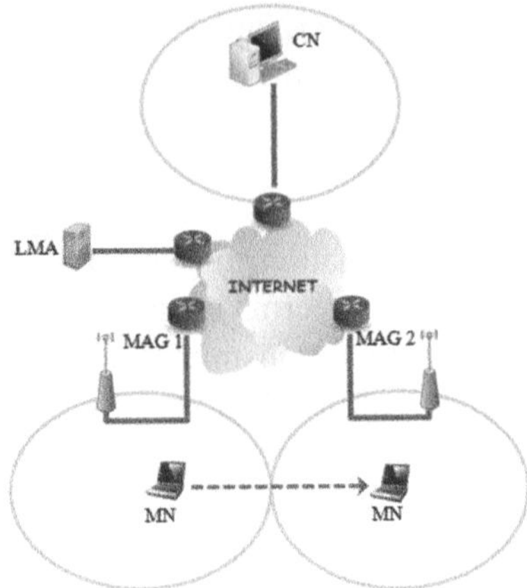

Figure 5.10. *PMIPv6 architecture*

When an MN arrives in a PMIPv6 domain (see Figure 5.11), it must first be attached to the access network in which it finds itself. The MAG1 can thus detect the arrival of this MN and determine using its identifier or MN-Id (which may be the *network access identifier* or the MAC address) if it is authorized to use the mobility services or not. If it is, the MAG1 sends a *proxy binding update* (PBU) to the LMA, including the MN identifier. This then allocates an HNP to the MN, updates the association table and implements a bidirectional tunnel to proxy-CoA1 of MAG1, which is considered as being the current CoA of the MN. The LMA then responds to the MAG1 with a *proxy binding acknowledgment* (PBA) including the MN HNP. The MAG1 also carries out the mechanisms necessary for the implementation of the tunnel to the LMA address (LMAA) and for the transfer of packets to the MN. The MAG1 sends an RA in which it indicates the HNP to the MN, the IPv6 prefix which the MN must use to configure its HoA. From then, all the packets destined for this prefix are routed by the LMA toward the MAG1 which, in turn, sends them to the MN. The packets sent by the MN will follow the same route.

When the MN changes network and therefore MAG attachment, the MAG1 detects the disconnection and launches a deregistration procedure with the LMA using a PBU/PBA exchange. The LMA then launches a timer at the end of which it deletes the association between the HNP and the proxy-CoA1. Once the MN is attached to the MAG2, in the same way as previously specified, a bidirectional tunnel between the LMA and the MAG2 implemented and all the communications from and to the MN go via this new tunnel.

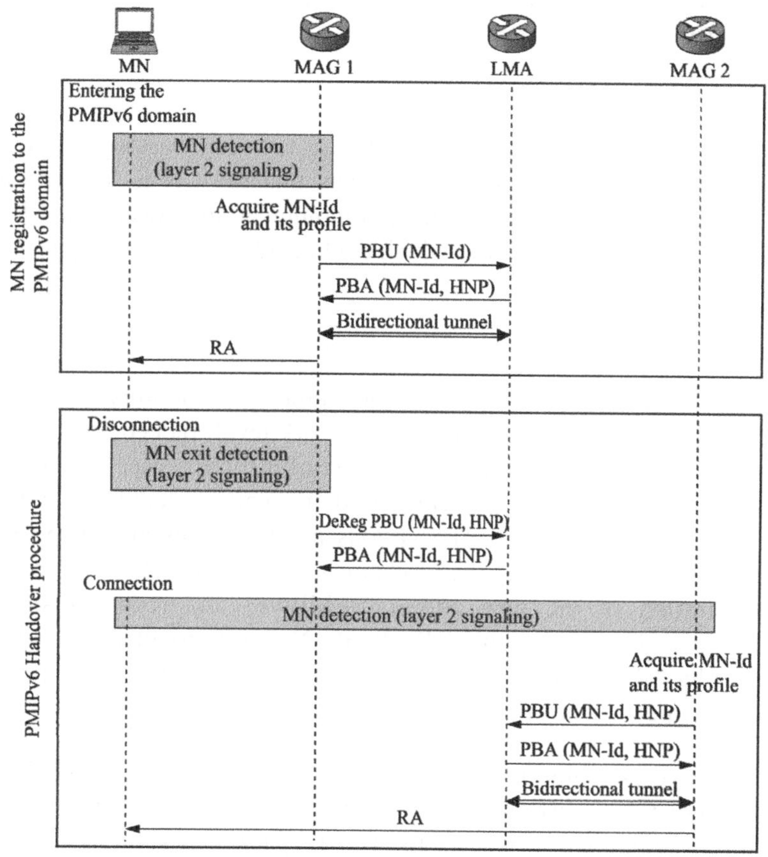

Figure 5.11. *Entry of an MN in a PMIPv6 domain and hand-over procedure*

It should be noted that in Figure 5.11, the deregistration at MAG1 is carried out before the registration at MAG2, but these two procedures are not correlated and can take place at the same time, or even in the other order depending on the time when the MAGs detect the connection/disconnection.

5.3.2.4.2. Management of multiple interfaces

The PMIPv6 standard provides the possibility for the MN to use several interfaces simultaneously. In this case, the LMA must allocate a mobility session (and one or more HNPs) to each of the interfaces. In the case of a hand-over between two interfaces, the LMA decides when it is necessary to create a new session and when it is necessary to update a new session depending on the parameters present in the PBUs (access technology used and hand-off indicator).

5.3.2.4.3. Routing optimization

PMIPv6 also sets out the possibility of avoiding passing through the MAG-LMA tunnel when the MAG detects that the CN is located on the same link as the MN. However, direct communication between the MN and CN is not set out in the standard, which could lead to long delays just as with Mobile IPv6.

5.3.2.5. *Conclusion on network level mobility*

As demonstrated previously, a large number of mobility management protocols have been put forward at the network level. Indeed, its position in the layered model enables the conservation of a mobile user's connectivity regardless of the technology being used as well as making hand-overs transparent for higher layers. The various improvements also help to limit hand-over time by enhancing the mechanisms for the detection of movement and registration of the new address (FMIPv6) and reducing signaling mobility to reduced domains (HMIPv6). Network-oriented management (PMIPv6) has also been proposed in order to avoid the MNs implementing the mobility mechanisms themselves.

However, these solutions suffer from some disadvantages, including the need to modify the network infrastructure (HA, PAR, NAR, MAP, MAG and LMA), adding extra headers to each IP packet as soon as the MN is located in a visited network, and the use of a bidirectional tunnel phase which is detrimental in terms of delaying the packet routing and interruption time.

5.3.3. *Mobility management with session initiation protocols (SIPs)*

One of the basic functionalities of the SIP protocol, presented in Chapter 2 and used as a foundation of IMS 3GPP architecture, is to manage the nomadic mobility of a mobile SIP client (mobility without communication in progress). This is carried out directly by redirection servers (see Figure 5.12).

Here, we will take the example of an MN moving away from its home network (1.domaine.fr) to a visited network (2.domaine.fr) without an SIP session in progress. Once this movement has been made, the MN must then be reregistered with its proxy (responsible for all the subdomains of domaine.fr) to update its location. After this movement, the CN attempts to reach the MN and thus sends it an INVITE request to the URI sip:mn@1.domaine.fr. The redirection server then responds with a standard "302 Moved Temporarily" message indicating another URI at which the CN can be reached. In our example, it indicates the URI sip:mn@2.domaine.fr. The CN, which receives the "302 Moved Temporarily" message, then traditionally exchanges INVITE/200 OK/ACK messages by using the new uniform resource identifier (URI) and the session can begin.

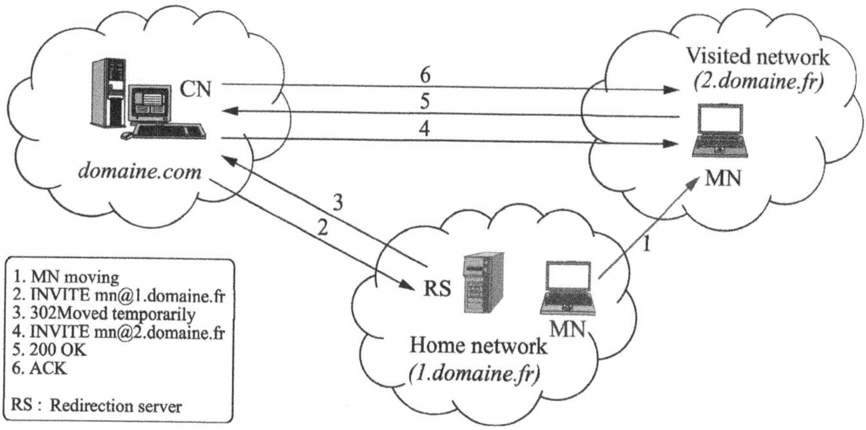

Figure 5.12. *SIP management of nomadic mobility*

After movement has taken place, this solution assumes that the MN must only be reregistered to update its location with the SIP proxy (and its redirection server) responsible for its home network, but in the event where the MN moves toward a domain under the responsibility of another SIP proxy, it can also be registered with this one. The operating mode of this reregistration is not specified in the standard in order to leave its implementation open.

5.3.3.1. *Continuous mobility management*

Although nomadic mobility management is an important asset offered by SIP mobility, the most interesting aspect of it in this chapter is the management of SIP mobility (while communication is in progress). Therefore, with SIP, a change of network while the SIP session remains active (e.g. a VoIP session) is possible.

In order to resolve this problem, there is a solution based on sending a re-INVITE message via an MN which carried out a change of network during the session. This re-INVITE message has exactly the same format as a traditional INVITE message and has the same Call-Id as the initial INVITE message. Nonetheless, the MN address must be modified at the level of the "contact" field as well as in the session description so that the session can restart from its new location.

The description of the session may also change in terms of the types of media to be sent, the codecs to be used, etc., which is one of the advantages of mobility management at the application level. Indeed, communication can adapt to the support level by negotiating new parameters, unlike lower level solutions which are transparent for applications.

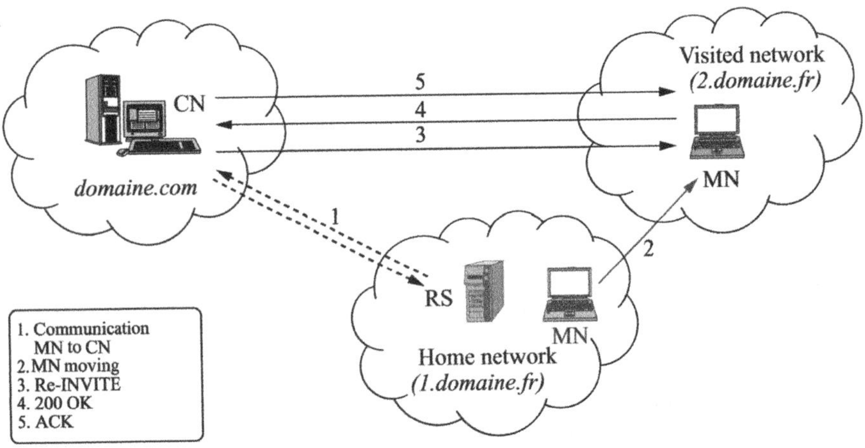

Figure 5.13. *SIP management by continuous mobility*

The mobility solution based on SIP helps to avoid the use of triangular routing, only requiring the same infrastructure as a traditional SIP architecture and works with IPv4 or IPv6. Moreover, the localization of users is a basic functionality of this protocol and SIP also helps to manage session, personal and service mobility.

5.3.3.2. *Limitations*

However, just like with the transport-level mobility solutions, mobility based on SIPs only helps to manage the applications controlled by SIP itself. However, this solution offers the advantage of transporting information which can be very useful in the management of QoS for applications with high time constraints which it helps initiate. However, it must be combined with another solution to allow mobility management for all types of applications.

This notion is discussed in the next section as a solution for mobility with guaranteed QoS in an IMS context.

5.4. Implementation of mobility solutions in hybrid systems

This section examines the integration of global mobility solutions into a satellite system. Indeed, the 250 ms delay introduced by the satellite could considerably extend the interruption time during a change of network. However, if the satellite systems truly wish to integrate with the Internet in the future, it is important for them to enable the implementation of the same services as terrestrial ones, and mobility is one of the most essential.

First, the objective is to evaluate the most promising mobility solutions in the framework of a DVB-S2/RCS satellite system. We will focus on MobileIPv6, FMIPv6, HMIPv6 and SIP mobility solutions which have the most chance of actually being implemented in the future. Indeed, MobileIPv6 and its extensions offer the advantage of adapting to all types of applications and SIP is the preferred multimedia session signaling protocol for most next-generation network (NGN) architectures.

Nevertheless, we have seen that SIP mobility is not fully specified and the registration procedure, for example, can be carried out in a number of different ways. Therefore, we will start by specifying the operation of SIP mobility integrated into a satellite system.

5.4.1. *Specification of SIP mobility in a DVB-S2/RCS system*

SIP mobility can be looked at in a number of different ways in a satellite system. Indeed, the satellite system can be considered as constituting a single and all-inclusive domain. In this case, a single proxy would be needed and would obviously be located either at the GW level for a star topology, or at the NCC level for a meshed topology. In a star topology, this would add no extra delay since all communications have to pass through the GW, but in a meshed topology each SIP message should go via the NCC, which would add at least one extra satellite hop per SIP message. This may be acceptable for the implementation of communication, but in terms of managing mobility, the interruption time would be too long to be taken into consideration.

Therefore, we would recommend a distributed solution in which an SIP proxy is deployed at the level of each ST. This would enable it to be compatible with both meshed and star topologies, and would reduce delays in establishing sessions and interruptions if communication is in progress. Furthermore, with a view to integrating a solution which combines QoS and mobility, this choice would be compatible with NGN architectures.

The following sections will look in detail at the implementation choices regarding the reregistration and reinitiation of sessions with a mobile SIP client, with a consideration of the various possibilities available.

5.4.1.1. *Problem of SIP reregistration with an MN*

If we consider an MN, initially registered at the level of its home SIP proxy, which moves between different networks of the satellite system, in any case a reregistration must be carried out at the level of the two SIP proxies: the new SIP proxy responsible for the domain which the MN has moved to and the home SIP proxy which may, therefore, redirect the SIP requests to the MN to the new SIP proxy.

Moreover, if the MN moves from a first visited network to a second visited network (both networks being distinct from the home network), the MN must deregister itself with the proxy in charge of the first visited network.

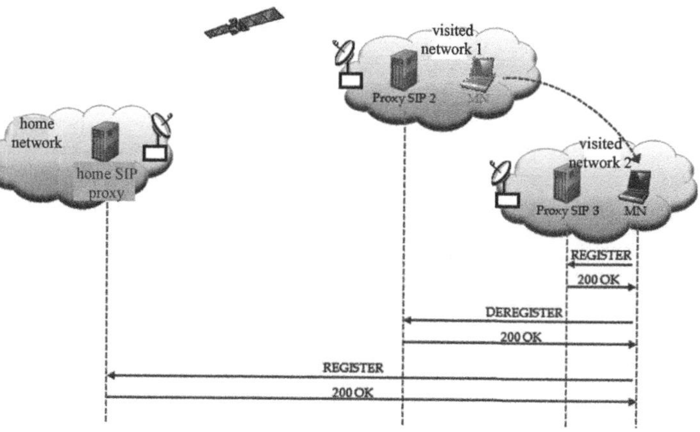

Figure 5.14. *Registrations initiated by the MN*

Given that the standard is vague on this point, there are several solutions which take account of the specific restrictions of the satellite:

– sending a REGISTER/DEREGISTER message for each SIP proxy involved (see Figure 5.14). Since these messages can be sent simultaneously from the MN to save time, the registration time is a minimum of two satellite hops, or 600 ms (it can be considered that one satellite hop = propagation time over the satellite link [250 ms] + propagation time over the WiFi and/or Ethernet connection + processing time at the emitter, receiver = 350 ms). However, this approach requires almost three identical messages to be sent over the radio link to which the MN has just connected, which is inefficient in terms of bandwidth use;

– sending a REGISTER message to the home SIP proxy which is responsible for transmitting the other message(s) to the other proxies involved (Figure 5.15). This solution resolves the problem of sending three messages via the radio link but poses a problem in terms of registration time. Indeed, four passages via the satellite link are necessary assuming that the REGISTER/DEREGISTER messages from the "Home Proxy SIP" are sent simultaneously. In the case of nomadic mobility, this approach is acceptable, but for continuous mobility the interruption time would be increased by at least 1.2 s (300 ms *4), which cannot be accepted. Moreover, this solution involves six SIP messages via the satellite link.

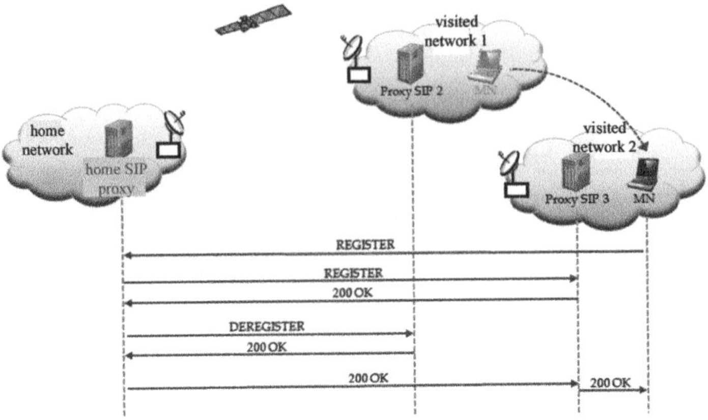

Figure 5.15. *Registrations initiated by the home SIP proxy*

– sending a REGISTER message to the SIP proxy responsible for the current domain containing the MN, and this SIP proxy then transmits the REGISTER/DEREGISTER message(s) to the other proxies (see Figure 5.16). This solution enables the optimal use of the radio link (two messages) as well as the satellite link (four messages). Moreover, it allows a registration time of 600 ms minimum.

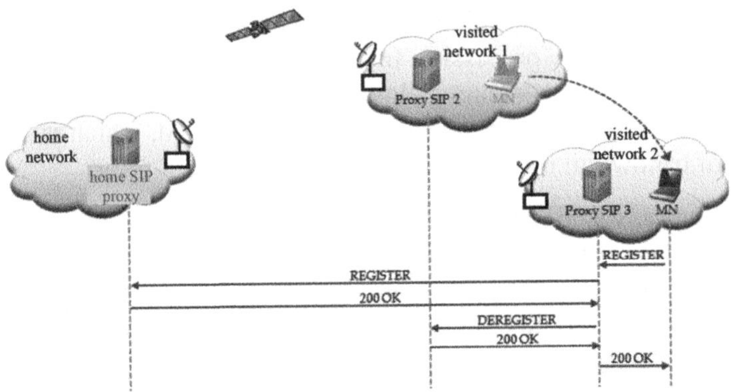

Figure 5.16. *Registrations initiated by the local SIP proxy*

The third solution involves registrations initiated by the local SIP proxy and is the most optimal. However, a registration delay of 600 ms remains detrimental during movement while communication is in progress. In this case, we will thus use a solution based on the third proposal, but with the final OK being sent back by the local SIP proxy to the MN, just after it receives the first REGISTER, as shown in Figure 5.17.

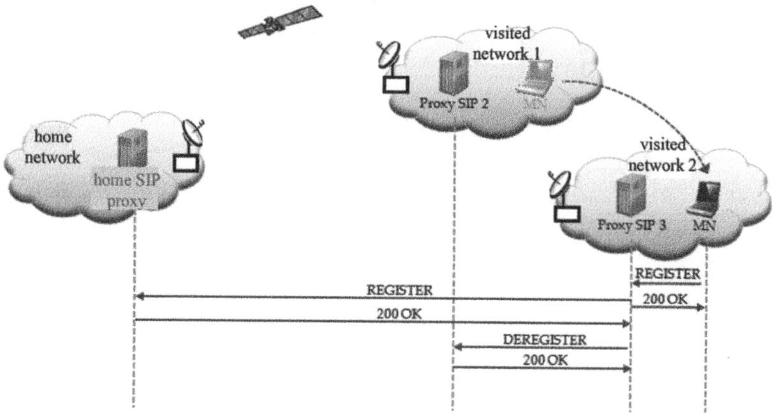

Figure 5.17. *Solution chosen for the SIP reregistration of an MN*

This solution (Figure 5.17) is quicker but does not guarantee that the reregistration with the home network and the deregistration with the previous proxy have actually taken place. This may be acceptable if the network does not offer a guaranteed QoS, but instead a best effort service. Inversely, in Figure 5.16, the reregistration procedure takes longer to carry out, but it guarantees that reregistration with the home network and the deregistration with the previous proxy have actually taken place. This mode can be used without being detrimental in the case of nomadic mobility.

5.4.1.2. *Problem of reinitiating the IP session in the case of continuous mobility*

Once deregistration is completed, it is necessary to inform the correspondent of a change of network address. To do this, as shown in section 5.3.3.1, an SIP session in progress can be reinitialized (Re-INVITE) when a movement takes place.

However, this solution only considers the case where both correspondents relaunch communication by exchanging messages directly between each other, without specifying the case where all the messages must go via the SIP proxies.

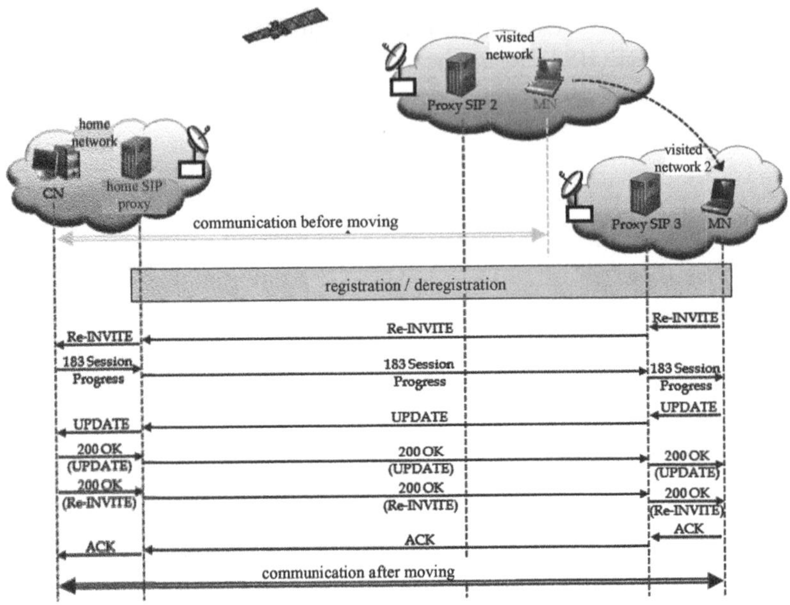

Figure 5.18. *Reinitiation of SIP session according to [RFC 3312]*

Figure 5.18 shows the process of messages necessary for the reinitialization of an SIP session after movement as an MN session is in progress. It can be assumed that at least five satellite hops are necessary in this figure, assuming that 200 OK (UPDATE) and the 200 OK (re-INVITE) are sent almost simultaneously. This, therefore, corresponds to a minimum interruption time of 1,500 ms just for the exchange of SIP messages, given that in order to obtain the global interruption time, it is also necessary to take account of the level 2 rebinding time for the reacquisition of an IPv6 address, etc.

With regard to extensions of the SIP protocol for the reservation of resources [CAM 02] (see Chapter 2), it is difficult to enable continuous mobility except if the mobile terminal is equipped with two interfaces and the reinitiation of the session can be carried out from the second interface while communication continues on the first. If this is not the case, the use of traditional SIP messages of re-INVITE/OK/ACK enables a gain of 600 ms, which is much more suitable in the case of an MN.

The authors of SIP even advocate a reinitialization solely based on re-INVITE and OK messages, without using the acknowledgement (ACK) which is essentially designed to make the reception of the OK more reliable. This, therefore, produces a gain of an extra 300 ms for the interruption time which corresponds to the exchange of SIP messages. In fact, if it is assumed that the CN sends messages just after the 200 OK, the MN will even receive its first packets 600 ms earlier compared to the case where the ACK is used (because in this case, the CN will send its first packets upon the reception of the ACK, which is two satellite hops later).

We will now focus on a solution which is compatible with the standard, assuming that the reinitiation of the SIP session will be carried out by the exchange of re-INVITE/200 OK/ACK messages as shown in Figure 5.19, and that if the QoS needs to be implemented, it will be at the OK level. However, in the case of a traditional SIP session initialization, the [CAM 02] may be followed. Nevertheless, by way of comparison, we will consider the case where the ACK continues to be exchanged but where communication can restart just after OK.

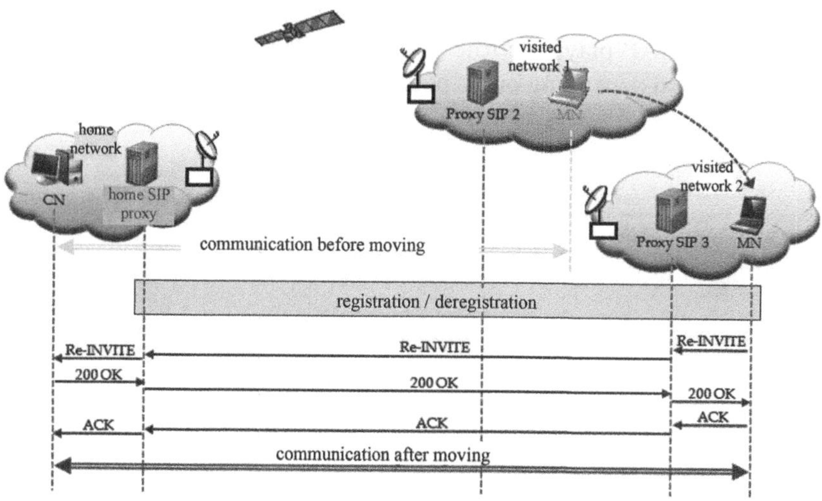

Figure 5.19. *Solution chosen for the reinitiation of the SIP session*

5.4.2. *Theoretical evaluations and recommendations*

This section will use theory to evaluate the interruption time and the delay in packet transfer for each of the mobility solutions proposed (Mobile

IPv6, HMIPv6, FMIPv6 and Mobility SIP). This is done for different types of movement, including the three shown in Figure 5.20. In this process, we will assume that the satellite system is made up of the following elements:

– a satellite in regenerative mode;

– three STs/GW fitted with a DVB-RCS return channel and communicating directly with each other. Moreover, the GW/ST1 connects the satellite system to the rest of the Internet;

– behind each ST/GW is an SIP proxy for SIP mobility and an MAP for HMIPv6;

– the home network is connected to the satellite system via the GW/ST1. It consists of an HA for MobileIPv6 and its extensions;

– the corresponding network is connected to the satellite system via the ST3. It contains of the CN;

– the host networks 1 and 2 are connected to the satellite system via the ST2;

– all the networks in question (home, correspondent and visited) are WiFi networks whose AP plays the role of AR and therefore the role of PAR or NAR for FMIPv6 depending on its position in the movement in question.

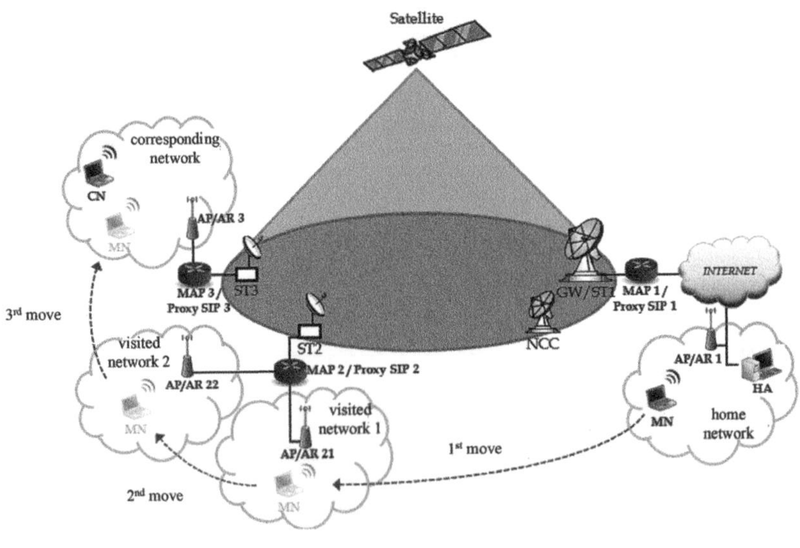

Figure 5.20. *The main types of movement in a satellite system*

The MN is initially located in its home network and it is assumed that it is in VoIP communication with the CN, since the session is controlled by SIP. Four types of movement are, therefore, taken into account (but only three appear in Figure 5.20 for reasons of clarity):

– movement 1 in which the MN goes via the home network to the visited network 1. This type of movement is considered as global mobility (or macromobility);

– movement 2 in which the MN goes from visited network 1 to visited network 2. This type of movement is considered as local mobility (or micromobility);

– movement 3 in which the MN goes from visited network 2 to the correspondent network, also considered as global mobility;

– movement 4 in which the MN goes from correspondent network to home network, also considered as global mobility. This movement does not appear in Figure 5.20.

Regarding interruption time T, this equates to the time taken between the last packet being received by the MN in the previous network and the first packet received by the MN in the new network. To calculate this, four times are used:

– *T2* represents the binding time to the new network. This time depends on the technology used and we will take an average value of T2 = 100 ms (not shown in the comparison Tables 5.1–5.5);

– *T3* represents the time necessary to obtain a new IPv6 address. Therefore, this is the time taken to receive an RA and the time needed for the DAD mechanisms. By default, the DAD time is 1,500 ms, but there is a version known as optimistic DAD which enables a use of the address before the termination of the procedure. This time equates to T3 = 1,525 ms or with optimistic DAD (DAD carried out later) T3 = 25 ms (by default T3 = 1,525 ms in comparison tables);

– *Tm1* represents the time taken between the sending of the first mobility protocol message (in the destination network) and the first packet received by the MN in the new network;

– *Tm2* represents the time taken between sending the first mobility protocol message (in the destination network) and the first packet received in "optimized route" mode by the MN in the new network. This then gives

Tm2 = Tm1 + "time taken between sending the first message necessary for the optimization of the route and the reception by the MN of the first packet in 'optimized route' mode".

Furthermore, transit time for the satellite network is estimated to be 300 ms, while transit time in the local network is negligible. These estimations are necessary to simplify the comparison between the mechanisms.

Finally, by way of a comparison, we will use the following values:

– *T** = time taken between the last packet received by the MN in the previous network and the first packet received in the optimized route by the MN in the new network. The following result is thus obtained: *T** = *T2* + *T3* + *Tm2*;

– *T'* = time taken between the last packet being received by the MN in the previous network and the first packet received by the tunnel between the HA and MN (and not by the FMIPv6 tunnel between the PAR and NAR). This notation is specific to FMIPv6;

– *D* = delay in transfer of packets received by the MN between T and T* (or for FMIPv6 between T and T' when T' exists);

– *D'* = delay in transfer of packets received by the MN between T' and T*;

– *D** = delay in transfer of packets received by the MN after T*.

When it is indicated that a delay or a time is ≈ 0, this indicates that it is very low, around several milliseconds at most.

We will now analyze in detail the results for each protocol separately.

5.4.2.1. *Mobile IPv6*

Regarding Mobile IPv6, it can be noted that the results are divided into three rows. The first row shows the use of Mobile IPv6 without route optimization (RO) or the RRT. In this case, when it moves outside the home network, the MN must receive an RA, continue to the DAD and then exchange a BU/BACK with its HA. The mobility of MN is thus fully transparent for the CN which continues to send packets to the home address of the MN. The packets are, therefore, intercepted by the HA which channels them to the current position of the MN. However, the delay in the transfer of the packets is still 600 ms, even when the MN and CN are behind the same ST (after movement 3) which seems highly unsuitable. This delay is, therefore, incompatible with audio and video conversations. Consequently, it is highly advisable to use MobileIPv6 with RO in the case of movement involving a satellite system.

	1st Move		2nd Move		3rd Move		4th Move	
	Interruption time	*Delay after handover*	*Interruption time*	*Delay after handover*	*Interruption time*	*Delay after handover*	*Interruption time*	*Delay after handover*
Mobile IPv6 no RO with DAD	Tm1 = 600 ms T = 2225 ms	D = 600 ms	Tm1 = 600 ms T = 2225 ms	D = 600 ms	Tm1 = 600 ms T = 2225 ms	D = 600 ms	T3 = 25 ms Tm1 ≈ 0 ms T = 125 ms	D = 300 ms
Mobile IPv6 with RO and RRT with DAD	Tm1 = 600 ms Tm2 = 2400 ms T = 2225 ms T* = 4025 ms	D = 600 ms D* = 300 ms	Tm1 = 600 ms Tm2 = 2400 ms T = 4025 ms T* = 4025 ms	D = 300 ms D* = 300 ms	Tm1 = 600 ms Tm2 = 1800 ms T = 3425 ms T* = 3425 ms	D ≈ 0 ms D* ≈ 0 ms	T3 = 25 ms Tm1 ≈ 0 ms Tm2 = 1200 ms T = 1325 ms T* = 1325 ms	D = 300 ms D* = 300 ms
Mobile IPv6 with RO and RRT with Optimistic DAD	T3 = 25 ms Tm1 = 600 ms Tm2 = 2400 ms T = 725 ms T* = 2525 ms	D = 600 ms D* = 300 ms	T3 = 25 ms Tm1 = 600 ms Tm2 = 2400 ms T = 2525 ms T* = 2525 ms	D = 300 ms D* = 300 ms	T3 = 25 ms Tm1 = 600 ms Tm2 = 1800 ms T = 1925 ms T* = 1925 ms	D ≈ 0 ms D* ≈ 0 ms	T3 = 25 ms Tm1 ≈ 0 ms Tm2 = 1200 ms T = 1325 ms T* = 1325 ms	D = 300 ms D* = 300 ms

Table 5.1. *Evaluations regarding Mobile IPv6*

In the event where RO is used (second row), the RRT procedure is considered as compulsory for the security reasons already mentioned. The interruption time differs depending on the type of movement considered:

– after movement 1, the MN must receive an RA (25 ms), proceed to the DAD (1,500 ms), exchange an initial BU/BACK with its HA (600 ms), carry out the RRT (home test (init) (HoTi/HoT) and care-of test (init) (CoTi/CoT) in parallel: 1,200 ms) and then exchange a second BU/BACK with the CN (600 ms). This procedure is, therefore, very time-consuming. However, the MN starts once again to receive data via the tunnel after implementing its association at the HA level, but these data are received with a 600 ms delay;

– after movement 2, the MN must carry out the same procedures as before, except that the CN and MN communicate directly while the MN is in visited network 1. This means that the BU at the HA level does not enable the implementation of a tunnel because the CN continues to send packets to the CoA obtained in visited network 1 and does not accept packets coming from a different address from this CoA without a new BU/BACK update. For communication to restart, the MN must, therefore, carry out the RRT and the update at the level of the CN. However, the communication will restart directly in "optimized route" mode, which means that $T=T^*$;

– after movement 3, management is similar to that in movement 2 except that the MN is in the same network as the CN. Hence, the exchanges between these two entities do not require a satellite hop and the interruption time is thus reduced. For the same reason, the delay in the transfer of packets is very low (≈ 0);

– regarding movement 4, the DAD procedure is not necessary since the MN uses its HoA. For the RRT procedure, only the HoTi/HoT exchange is necessary here since the CoA corresponds with the HoA of the MN. The interruption time is, therefore, much greater in the case of MobileIPv6 without RO since communication cannot restart before the MN updates its binding at the CN level (HoTi/HoT and then BU/BACK).

5.4.2.2. *HMIPv6*

For HMIPv6, we will now re-examine the same three cases as for Mobile IPv6. We can assume that when the MN arrives in the new network, the time needed for it to obtain an RCoA and an LCoA is the same as obtaining a CoA in the case of Mobile IPv6 (it can be assumed that the two procedures are carried out in parallel).

	1st Move		2nd Move		3rd Move		4th Move	
	Interruption time	*Delay after handover*	*Interruption time*	*Delay after handover*	*Interruption time*	*Delay after handover*	*Interruption time*	*Delay after handover*
HMIPv6 no RO with DAD	Tm1 = 600 ms T = 2225 ms	D = 600 ms	Tm1 ≈ 0 ms T = 1625 ms	D = 600 ms	Tm1 = 600 ms T = 2225 ms	D = 600 ms	Tm1 ≈ 0 ms T = 1625 ms	D = 300 ms
HMIPv6 with RO and RRT with DAD	Tm1 = 600 ms Tm2 = 2400 ms T = 2225 ms T* = 4025 ms	D = 600 ms D* = 300 ms	Tm1 ≈ 0 ms Tm2 = 0 ms T = 1625 ms T* = 1625 ms	D = 300 ms D* = 300 ms	Tm1 = 600 ms Tm2 = 1800 ms T = 3425 ms T* = 3425 ms	D ≈ 0 ms D* ≈ 0 ms	Tm1 ≈ 0 ms Tm2 = 1200 ms T = 2825 ms T* = 2825 ms	D = 300 ms D* = 300 ms
HMIPv6 with RO and RRT with Optimistic DAD	T3 = 25 ms Tm1 = 600 ms Tm2 = 2400 ms T = 725 ms T* = 2525 ms	D = 600 ms D* = 300 ms	T3 = 25 ms Tm1 ≈ 0 ms Tm2 = 0 ms T = 125 ms T* = 125 ms	D = 300 ms D* = 300 ms	T3 = 25 ms Tm1 = 600 ms Tm2 = 1800 ms T = 1925 ms T* = 1925 ms	D ≈ 0 ms D* ≈ 0 ms	T3 = 25 ms Tm1 ≈ 0 ms Tm2 = 1200 ms T = 1325 ms T* = 1325 ms	D = 300 ms D* = 300 ms

Table 5.2. *Evaluations regarding HMIPv6*

When RO is not used, the same problem arises as it was seen with Mobile IPv6, meaning a delay in the transfer of packets between the CN and MN of around 600 ms as soon as the MN is outside the home network. However, it can be noted that for movement 2, the time T is significantly reduced since the MN simply needs to inform its MAP locally (hence, a very low Tm1). Inversely, in the case of returning to the home network (movement 4), unlike with Mobile IPv6, the MN must obtain an RCoA and an LCoA, and the interruption time is thus extended by 1,500 ms (DAD procedure). It is also important to note that in the case of HMIPv6, communication is made using multiple encapsulations since the packets sent from the CN to the HoA of the MN are first encapsulated by the HA going to the MAP which encapsulates them once again as they go to the LCoA of the MN. This, therefore, produces a significant overhead, which creates an inefficient use of the satellite and WiFi links.

When the RO is activated (and therefore RRT), it can be noted that the results obtained are similar to those achieved by Mobile IPv6 in the case of movements 1 and 3. However, the case of movement 2 which is a case of micromobility shows the real advantage of the HMIPv6 protocol: no message needs to pass through the satellite, therefore, the interruption time is reduced to T2+T3 and the delay in the transfer of packets remains lower than the recommendations advocated for VoIP or video-conference applications (< 400 ms). For movement 4, the DAD procedure must also be carried out, unlike for Mobile IPv6.

If the optimistic DAD method is used, a considerable time saving can once again be observed, especially in the case of movement 2 where the interruption time is only 125 ms.

5.4.2.3. *FMIPv6 in predictive mode*

In predictive mode, it is assumed that the movement of the MN is slow enough for it to be in the common space between the departure and arrival networks long enough to be able to carry out the procedure necessary before the change of network is completed. In the case where the DAD procedure is used, this transition time must, therefore, be at least greater than the time necessary to exchange RtSolPr/PrRtAdv, FBU, HI/HACK and FBACK messages as well as the time necessary for DAD mechanisms to operate. This equates to a time greater than two satellite hops (since HI and HACK messages cross the satellite) + 1,500 ms, thus greater than 2,100 ms overall. If the DAD procedure is not carried out, this time is reduced to 600 ms. These different hypotheses are the best-case scenarios for FMIPv6.

As for the previous protocols, we start by considering the case of using FMIPv6 without RO and with the DAD procedure. In this case, if the previous conditions are combined, the interruption time is still reduced to T2+T3 = 125 ms (indeed, assuming that for the MN to know that it has arrived in the new network, it must receive an RA, hence T3 = 25 ms). For the first three movements, the tunnel between the PAR and NAR must be maintained, at least for T'–T (= 600 ms). This is the time for the MN to inform its HA of the new address. However, after each of these three movements, the delay in the transfer of packets between the MN and CN is around 600 ms, because the CN continues to send to the HoA of the MN, making it incompatible with the recommendations on the delay for VoIP and video-conferencing. In the case of movement 4, the PAR-NAR tunnel must be maintained only while the MN exchanges BU/BACK messages locally with its HA and the delay D falls to 300 ms. Tm1 is low (≈ 0 ms) because this is the time taken between the UNA being sent and the first packet being received by the PAR-NAR tunnel, which are exchanged locally.

	1st Move		2nd Move		3rd Move		4th Move	
	Delay after handover	*Interruption time*	*Delay after handover*	*Interruption time*	*Delay after handover*	*Interruption time*	*Delay after handover*	*Interruption time*
Predictive FMIPv6 no RO with DAD	T3 = 25 ms Tm1 ≈0 ms T = 125 ms T' = 725 ms	D = 600 ms D' = 600 ms	T3 = 25 ms Tm1 ≈0 ms T = 125 ms T' = 725 ms	D = 600 ms D' = 600 ms	T3 = 25 ms Tm1 ≈0 ms T = 125 ms T' = 725 ms	D = 600 ms D' = 600 ms	T3 = 25 ms Tm1 ≈0 ms T = 125 ms T' = 125 ms	D = 300 ms D' = 300 ms
Predictive FMIPv6 with RO and RRT with DAD	T3 = 25 ms Tm1 ≈0 ms Tm2 = 2400 ms T = 125 ms T' = 725 ms T* = 2525 ms	D = 600 ms D' = 600 ms D* = 300 ms	T3 = 25 ms Tm1 ≈0 ms Tm2 = 2400 ms T = 125 ms T* = 2525 ms	D = 300 ms D* = 300 ms	T3 = 25 ms Tm1 ≈0 ms Tm2 = 1800 ms T = 125 ms T* = 1925 ms	D = 600 ms D* ≈0 ms	T3 = 25 ms Tm1 ≈0 ms Tm2 = 1200 ms T = 125 ms T* = 1325 ms	D = 300 ms D* = 300 ms

Table 5.3. *Evaluations regarding FMIPv6 in predictive mode*

When the RO and RRT procedures are carried out, they reduce the delay, for movements 1–3, to D* = 300 ms. However, the most interesting case is in movement 2, whose interruption time T is only 125 ms and a delay D of 300 ms. The case featuring movement 4 also delivers a high level of performance. Values T' and D' do not occur for the final three movements because, after a movement of the MN, the CN continues to send its packets to the previous CoA of the MN, which are then intercepted by the PAR and channeled toward the NAR. The exchange of a BU/BACK message with the HA changes nothing. The PAR-NAR tunnel must, therefore, be maintained for at least T*– T.

5.4.2.4. *FMIPv6 in reactive mode*

Here, we examine the worst of the cases involving FMIPv6 where the MN does not have time to send an FBU to its PAR. The DAD procedure is, therefore, carried out upon reception of the HI. Moreover, the PAR-NAR tunnel can only be implemented after the exchange of FBU/HI/HACK/FBACK messages which, apart from the case of movement 2, must all go via the satellite.

The case of movement 1 without RO and with DAD first highlights the huge difference between the interruption time for FMIPv6 in reactive mode (2,845 ms) and predictive mode (125 ms). Unlike with the predictive mode, in the reactive case, the PAR-NAR tunnel must be implemented while the communication is interrupted. Furthermore, for movements 1, 2 and 3, we find the same problem as in the previous cases without RO: the delay is 600 ms because the CN continues to send to the HoA of the MN. For these movements, the PAR-NAR tunnel must also be maintained for the time it takes for the MN to inform its HA, which is at least 600 ms from T.

	1st Move		2nd Move		3rd Move		4th Move	
	Delay after handover	*Interruption time*	*Delay after handover*	*Interruption time*	*Delay after handover*	*Interruption time*	*Delay after handover*	*Interruption time*
Reactive FMIPv6 no RO with DAD	Tm1 = 1800 ms T = 2825 ms T' = 3425 ms	D = 600 ms D' = 600 ms	Tm1 = 600 ms T = 1625 ms T' = 2225 ms	D = 600 ms D' = 600 ms	Tm1 = 1800 ms T = 2825 ms T' = 3425 ms	D = 600 ms D' = 600 ms	T3 = 25 ms Tm1 = 1200 ms T = 1325 ms T' = 1325 ms	D = 300 ms D' = 300 ms
Reactive FMIPv6 with RO and RRT with DAD	Tm1 = 1800 ms Tm2 = 3600 ms T = 2825 ms T' = 3425 ms T* = 5225 ms	D = 600 ms D' = 600 ms D* = 300 ms	Tm1 = 600 ms Tm2 = 2400 ms T = 1625 ms T* = 4025 ms	D = 600 ms D* = 300 ms	Tm1 = 1800 ms Tm2 = 3000 ms T = 2825 ms T* = 4625 ms	D = 600 ms D* ≈0 ms	T3 = 25 ms Tm1 = 1200 ms Tm2 = 2400 ms T = 1325 ms T* = 2525 ms	D = 300 ms D* = 300 ms
Reactive FMIPv6 with RO and RRT with Optimistic DAD	T3 = 25 ms Tm1 = 1800 ms Tm2 = 3600 ms T = 1325 ms T' = 1925 ms T* = 3725 ms	D = 600 ms D' = 600 ms D* = 300 ms	T3 = 25 ms Tm1 = 600 ms Tm2 = 2400 ms T = 125 ms T* = 2525 ms	D = 600 ms D* = 300 ms	T3 = 25 ms Tm1 = 1800 ms Tm2 = 3000 ms T = 1325 ms T* = 3125 ms	D = 600 ms D* ≈0 ms	T3 = 25 ms Tm1 = 1200 ms Tm2 = 2400 ms T = 1325 ms T* = 2525 ms	D = 300 ms D* = 300 ms

Table 5.4. *Evaluations regarding FMIPv6 in reactive mode*

When the RO and RRT procedures are activated, the delay in the transfer of packets between the CN and MN can be reduced to 300 ms from T*, but the time necessary for the packets to arrive via the optimized route is very long (up to 5.225 s for movement 1). In fact, this also shows that, even if FMIPv6 may appear to be the most efficient solution when predictive mode is possible, when this is not the case, this protocol can also become the worst possible option.

When the optimistic DAD procedure is possible, it provides a gain of 1.5 s for the first three movements, as for the previous protocols.

5.4.2.5. *SIP mobility*

This section examines the case of SIP mobility, assuming that communication can be relaunched either before the ACK is sent (after the OK), or after and for each of the cases, the DAD procedure or the optimistic DAD procedure can be used.

The main characteristic which stands out is that there is no pathway for a non-optimized phase where the delay is 600 ms, which means that for all the cases considered, T=T* and D=D*. It can also be observed that the use of the optimistic DAD procedure leads to precious time saving and helps to reduce by half the interruption time (in the case of movement 3, it is even reduced to around T2 + T3 = 125 ms).

Apart from movement 3, it can also be seen that the interruption time T (or T*) is reduced by 600 ms if SIP mobility with ACK is compared with SIP mobility without ACK.

	1st Move		2nd Move		3rd Move		4th Move	
	Delay after handover	*Interruption time*	*Delay after handover*	*Interruption time*	*Delay after handover*	*Interruption time*	*Delay after handover*	*Interruption time*
SIP mobility after ACK with DAD	Tm2= 1200 ms T = 2825 ms T' = 2825 ms	D = 300 ms D* = 300 ms	Tm2= 1200 ms T = 2825 ms T' = 2825 ms	D = 300 ms D* = 300 ms	Tm2 ≈ 0 ms T = 1625 ms T' = 1625 ms	D ≈ 0 ms D* ≈ 0 ms	Tm2= 1200 ms T = 2825 ms T' = 2825 ms	D = 300 ms D* = 300 ms
SIP mobility after ACK with Optimistic DAD	T3 = 25 ms Tm2 = 1200 ms T = 1325 ms T* = 1325 ms	D = 300 ms D* = 300 ms	T3 = 25 ms Tm2 = 1200 ms T = 1325 ms T* = 1325 ms	D = 300 ms D* = 300 ms	T3 = 25 ms Tm2 ≈ 0 ms T = 125 ms T* = 125 ms	D ≈ 0 ms D* ≈ 0 ms	T3 = 25 ms Tm2 = 1200 ms T = 1325 ms T* = 1325 ms	D = 300 ms D* = 300 ms
SIP mobility before ACK with DAD	Tm2 = 600 ms T = 2225 ms T* = 2225 ms	D = 300 ms D* = 300 ms	Tm2 = 600 ms T = 2225 ms T* = 2225 ms	D = 300 ms D* = 300 ms	Tm2 ≈ 0 ms T = 1625 ms T* = 1625 ms	D ≈ 0 ms D* ≈ 0 ms	Tm2 = 600 ms T = 2225 ms T* = 2225 ms	D = 300 ms D* = 300 ms
SIP mobility before ACK with Optimistic DAD	T3 = 25 ms Tm2 = 600 ms T = 725 ms T* = 725 ms	D = 300 ms D* = 300 ms	T3 = 25 ms Tm2 = 600 ms T = 725 ms T* = 725 ms	D = 300 ms D* = 300 ms	T3 = 25 ms Tm2 ≈ 0 ms T = 125 ms T* = 125 ms	D ≈ 0 ms D* ≈ 0 ms	T3 = 25 ms Tm2 = 600 ms T = 725 ms T* = 725 ms	D = 300 ms D* = 300 ms

Table 5.5. *Evaluations regarding SIP mobility*

5.4.2.6. *Comparison of solutions*

In this section, we will compare the various solutions proposed above for each type of movement. Two types of comparison are, therefore, necessary: one which examines solely the interruption time T, even if the first packets arrive with a delay of 600 ms, and the other which only looks at the time from when the delay is 300 ms and thus remains compatible with the ITU telecommunication standardization sector (ITU-T) recommendations on VoIP applications or videoconferencing.

For the first type of comparison, the case which provides the highest performance overall is FMIPv6 in predictive mode since it enables a reduction in time T = 125 ms. Another advantage of this solution is that it remains very effective even when using the DAD procedure since it is carried out in the new network while communication is still taking place in the previous network. However, the implementation of a temporary tunnel between PAR and NAR is always necessary and entails, for movements 1 and 3, a double satellite hop. With regard to movement 2, HMIPv6 with RO, RRT and optimistic DAD is the best solution since it also gives T = 125 ms with a delay D = 300 ms (like predictive FMIPv6) but moreover, the only messages which are necessary are the LBU/BACK exchanged with the MAP, unlike many messages necessary for the implementation of the FMIPv6 tunnel. The solution Mobile IPv6 without RO also enables an efficient management of movement 4, with T = 125 ms and D = 300 ms, without requiring any other messages apart from the binding update/binding acknowledgement (BU/BA) exchanged with the HA. However, the final two solutions are only efficient in the case of a very precise movement at the detriment of other types of movement.

For the second type of comparison, we will compare each type of movement in order only taking account of solutions with RO and RRT:

– for movement 1, if we compare solutions using the optimistic DAD procedure, it can be observed that the solutions based on SIP mobility are those which deliver the highest performance and help to reduce the interruption time by over 1 s in comparison with solutions based on Mobile IPv6;

– for movement 2, the HMIPv6 solutions (with optimistic DAD) and predictive FMIPv6 remain the most efficient, for the same reasons as for the first type of comparison;

– for movement 3, SIP mobility solutions are also the best option whether DAD or optimistic DAD mechanisms are used. Indeed, they help to re-establish communication directly between the MN and CN in around 125 ms when the optimistic DAD procedure is used;

– for movement 4, the solution of predictive FMIPv6 is the most efficient.

5.4.2.7. *Recommendations regarding mobility in a satellite system*

The first remark regarding Mobile IPv6 and its extensions is that, as part of a satellite system, it is essential that the RO and RRT procedures can be carried out in order to be compatible with the ITU-T recommendations for multimedia applications. Moreover, the bidirectional tunnel phase involves IPv6/IPv6 encapsulations which add a significant overhead to the system. The need to pass through the satellite system twice leads to a greater use of the system's overall resources, thus confirming the essential nature of the RO and RRT procedures.

The second important point regarding Mobile IPv6 and HMIPv6 with RO and RRT is that the bidirectional tunnel phase can only be used for a single type of movement: when the MN leaves the home network. This means that no application (including that with no restriction in terms of delay) can take advantage of the bidirectional tunnel phase for any of the other types of movement, which significantly reduces the value of this solution for a user moving between different visited networks.

It can also be noted that the time needed to carry out the DAD mechanisms is much too long. Mechanisms similar to optimistic DAD must, therefore, be implemented. Mobile IPv6 (with optimistic DAD) is as efficient as its extensions for movements 1 and 3.

FMIPv6 is very efficient for movements without a change of ST (movement type 2) and movements with an ST from the CN network to another network (movement type 4) as long as the predictive mode can be used. However, this solution is much less efficient for other types of movement mainly because RO and RRT procedures, necessary for reducing the delay to one satellite hop, take a long time to carry out. Furthermore, if the predictive mode cannot be carried out, the interruption times become highly prohibitive and are even higher than all those obtained with Mobile IPv6. FMIPv6 must, therefore, be used while carefully ensuring that the conditions are right for the predictive mode to be applied.

HMIPv6 is efficient for cases of micromobility (type 2 movement) for which, after all, it was designed. However, given the results for the other types of movement, it is essential for HMIPv6 to be combined with one or more other mobility protocols.

Finally, in terms of mobility management for multimedia applications based on SIP in a DVB-S2/RCS satellite system, SIP mobility is a particularly efficient solution. Indeed, it offers the huge advantage of being able to relaunch communication directly between the two entities in question, unlike the previous solutions, which therefore enables it to efficiently manage all types of movement without depending on a specific topology, and even to be the most efficient solution for movements 1 and 3. Moreover, not using ACK (or at least relaunching communication just after the OK) helps to further improve its performance. Finally, it adds no overhead when the MN is in a visited network.

SIP mobility is, therefore, a good alternative to Mobile IPv6 or its extensions when mobility management in a satellite system has significant time restrictions, due to its mobility-related performance, and also its direct connection with the QoS configuration. Nevertheless, Mobile IPv6 (and its extensions) remains an efficient solution for mobility management of applications with fewer restrictions, especially in terms of transmission delay.

5.5. SIP for mobility management and QoS for interactive applications

This section defines a solution which enables mobility and QoS management in a DVB-S2/RCS system solely using SIP, independently from any other mobility protocol. This solution would, therefore, allow SIP applications to implement QoS and manage network changes in a satellite system in an entirely automatic way. It is, therefore, more specifically suited to multimedia applications with high constraints. This solution is made even more interesting by the fact that it is compatible with the IMS architecture discussed in Chapter 4.

An SIP proxy, responsible for the reservation of resources (see Chapter 4), playing the role of see CSCF (P-CSCF), must be located behind each ST to be able to intercept SIP messages and transmit QoS information to the required elements (ST, policy decision function (PDF), NCC, etc.)

When the session is initiated, QoS reservations are made following the procedures described in Chapter 4, varying according to the configuration of the satellite system (transparent and meshed). The session reinitialization and resource reservation after a change of network are carried out as shown in Figure 5.19 (re-INVITE). For the traditional session ending (messages BYE/OK), the release of resources is triggered by the reception of a BYE message. However, for a change of network, no BYE message is exchanged with the previous SIP proxy. The only message sent to it is the REGISTER message (if the previous SIP proxy is also the home proxy) or DEREGISTER (for other cases). The release of resources must, therefore, be initiated at the moment that this message is received which indicates in any case that the MN is now located in another network. This new functionality must be added to the SIP proxy. Moreover, the exchange of REGISTER/DEREGISTER messages between the various SIP proxies is not a functionality which is described in the RFC and must, therefore, be added.

Finally, although the exchange of REGISTER/DEREGISTER messages enables the release of resources at the level of the previous ST responsible for the MN, this is not the case for the one responsible for the CN. The release of resources must, therefore, be carried out at the level of the session reinitialization. When the SIP proxy responsible for the CN receives a re-INVITE message with the same Call-Id as the one for the current session, it understands that this corresponds to a session modification and analyzes the address of the SIP client which sent this message:

– if the address corresponds to neither of the two SIP client addresses participating in the session, it understands that this corresponds to a session modification with a change of network and therefore releases the associated resources;

– if the address corresponds to one of the two SIP client addresses participating in the session, this corresponds to a session modification without a network change. Therefore, it does not release resources and waits to receive the OK message (or session progress) to modify the reservation. Indeed, in this case, the session is not interrupted and releasing resources could harm the quality of the communication until the re-reservation.

Subsequently, when the SIP proxy on the CN side detects that there will be a session modification with a network change, there are two possibilities when it receives the OK message:

– the MN has moved behind a different ST from that of the CN: in this case, the SIP proxy on the CN side carries out a rereservation of resources specifying the new address and possibly the new session parameters;

– THE MN has moved behind the same ST as that of the CN: in this case, the SIP proxy detects that the session will no longer pass through the system.

All these functionalities must, therefore, also be implemented at the level of the SIP proxy.

Looking at mobility from the user's side, an augmented SIP client must be implemented. Various functionalities must, therefore, be added:

– a network change detection module (and dynamic discovery of the local SIP proxy) interfaced with the SIP client to inform it that it must launch reregistration and session reinitiation procedures;

– sending a specific REGISTER message to indicate the need to inform other SIP proxies;

– update of session parameters to send a re-INVITE message containing the new MN address and the possible new session description protocol (SDP) parameters.

It should be noted that unlike Mobile IPv6, no major modification is necessary in this case at the level of the CN. The CN effectively considers a re-INVITE message as a traditional modification of the current session and therefore only needs to update the session parameters. The only modification necessary for this architecture is to indicate that in the case of a re-INVITE message, it must respond directly by sending an OK (INVITE) message without using messages such as 183 session progress, etc.

5.6. Evaluation of mobility solutions in a simulated DVB-S2/RCS architecture

In order to compare the performance of various protocols in a satellite/terrestrial hybrid network environment, Mobile IPv6, FMIPv6 (where relevant) and SIP mobility have been deployed and tested via a satellite network simulator (OpenSand[1]). This comparison mainly looks at the interruption times obtained for various cases of mobility involved in the

1 http://opensand.org.

framework of a DVB-S2/RCS satellite system with a mobile user during a multimedia session (VoIP or videoconference). However, we will also examine the problems of overheads for each solution.

5.6.1. *Comparison of interruption times*

The topology and types of movement shown in Figure 5.20 will be applied to the test bed in order to compare the various interruption times, but without movement 2, since HMIPv6 was not evaluated in this framework. WiFi was chosen as a level 2 technology for the terrestrial network. In order to reduce the rebinding time with the new AP as much as possible, the WiFi cards will be configured to work on a predefined channel for each AP, which would give level 2 interruption times of around 0.1–0.2 s. To be able to compare the various solutions in the fairest way possible, we will only take interruption times into account from our results when the level 2 time is located in this window. Likewise, in order to calculate the average interruption time obtained for each solution and each movement, first, we will only take account of the most common cases where mobility messages are exchanged as defined in section 5.4.2. We will then examine individual cases where interruption times are the highest or the lowest in the most common cases and attempt to explain the reasons for these results.

5.6.2. *Common cases*

The theoretical evaluation of solutions was focused on the case where the MN was the receiver. This evaluation examines the case of bidirectional multimedia communication. We will also look at interruption time in a case where the MN is the emitter, which is the time taken between the last packet being sent from the previous network (and received by the correspondent) and the first packet being sent from the new network (and received by the correspondent). Indeed, these two times can differ depending on the type of movement and the type of solution in question. Moreover, regarding the RO phase for Mobile IPv6, we will also add the case where the BACK message is not necessary between the MN and CN, since this option is allowed in the MIPL2.0.2 stack. This allows the MN to start sending packets again directly to the CN up to 600 ms faster when the MN and CN are located behind a different ST.

For interruption times corresponding to FMIPv6, the reception and sending of messages via the PAR-NAR tunnel could not be observed (due to the tunnel), but all the other FMIPv6 messages were able to be observed. Taking account of the observations carried out on Mobile IPv6, we can, therefore, consider that the first message sent from the new network is transmitted just after the sending of the UNA message (+ 0.05 s after) and that the first message received is sent 0.1 s after the sending of the UNA (similar previous experiments were used in order to establish these times).

Figures 5.21 and 5.22, therefore, show the average interruption times (20 measures for each movement and each solution) corresponding to the various solutions given, examining cases with DAD or optimistic DAD (activating the option CONFIG_IPV6_OPTIMISTIC_DAD) and only taking account of the times where the delay in the transmission of packets between the CN and MN is below 400 ms:

– SIP mobility with ACK (the CN sends data just after receiving the ACK);

– SIP mobility without ACK (the CN sends data just after sending the OK (INVITE));

– Mobile IPv6 with BACK (the MN sends data directly to the CN just after receiving the BACK sent by the CN);

– Mobile IPv6 without BACK (the MN sends data directly to the CN just after sending the BU to it);

– FMIPv6 in predictive mode, as for Mobile IPv6, with or without BACK messages being exchanged with the CN.

Nonetheless, in the case of Mobile IPv6 and FMIPv6, the interruption time experimented by the MN as a receiver is the same, whether there is a BACK message sent or not, since in any case the CN starts sending packets when it receives the BU sent by the MN.

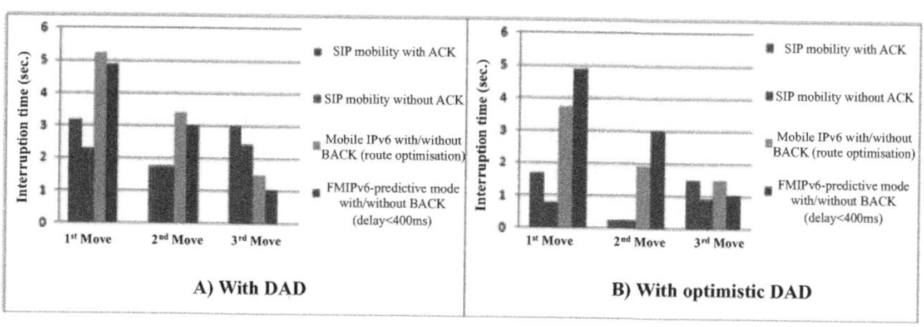

Figure 5.21. *Interruption times registered by the MN as a receiver*

Compared to the time theoretically calculated, the following significant differences can be noted:

– in the case of movement 1 for Mobile IPv6 and FMIPv6, the exchange of BU/BACK messages between the MN and HA takes 1 s longer than predicted, due to a glitch in the MIPL 2.0.2 stack;

– in the case of FMIPv6, the sending of the UNA message takes an average of 1.2 s, but varies between a minimum of 0.26 s and a maximum of 2.47 s, which considerably extends the interruption time and above all makes it very variable. In the case where the optimistic DAD method is used, FMIPv6 then becomes the solution with the longest interruption times for movements 1 and 2.

Let us now compare the solution for each movement:

– for movement 1, even taking account of the implementation flaws described previously, the solutions based on SIP are generally more efficient at both receiving and sending, especially when the optimistic DAD mechanisms are used. As an indication, in the case of FMIPv6, the UNA message is sent on average at t = 1.35 s for measures carried out for this movement. Likewise, for the Mobile IPv6 solution with DAD, the first message is sent by the bidirectional tunnel (only valid for this movement for the reasons given in section 5.4.2.1) at t = 3.36 s on average (but this time can also suffer from implementation flaws);

– for movement 3, SIP solutions are also the most efficient;

– for movement 4, SIP solutions are less efficient when the DAD is activated because the other solutions do not need to use these mechanisms when the MN reaches its home network, but this obstacle is overcome when optimistic DAD mechanisms are used.

Regarding Mobile IPv6 and FMIPv6, it can be observed that the option of deactivating the BACK message in the RO phase enables packets to be sent directly to the CN up to 600 ms faster. This option also enables the phase commonly known as “triangular routing” during which the MN sends packets directly to the CN while the CN continues to send packets via the HA. This phase takes place between the moment when the MN sends the BU to the CN and the moment when the CN receives the BU.

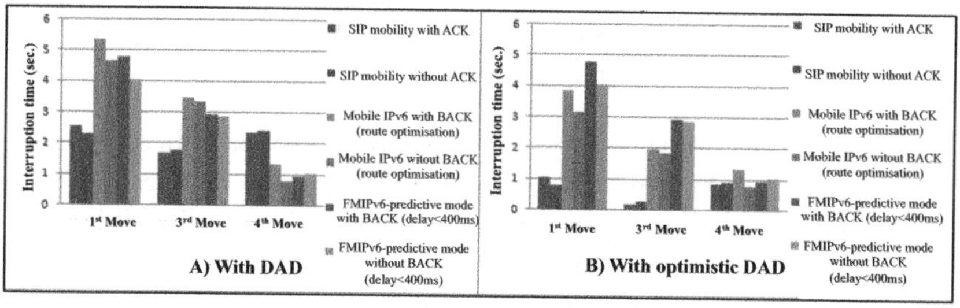

Figure 5.22. *Interruption time registered by the MN as emitter*

More generally, if we consider a mobile user during VoIP or videoconferencing communication, in the framework of a DVB-S2/RCS satellite system, we can conclude that:

– for Mobile IPv6 with RO and RRT, the bidirectional tunnel phase (only useable when the MN leaves its home network) does not comply with the ITU-T recommendations. The interruption time is, therefore, significantly increased as soon as the Mobile IPv6 messages need to cross the satellite system;

– for FMIPv6, the PAR-NAR tunnel only complies with the ITU-T recommendation when the MN leaves the network of the CN to any other network. The RO and RRT procedures specific to Mobile IPv6 are, therefore, necessary and detrimental in terms of interruption time;

– solutions based on SIP are more efficient in general, except for a return to the home network where DAD mechanisms, when activated, are detrimental. However, this problem can be overcome by the use of mechanisms such as optimistic DAD. Indeed, communication can restart directly between the CN and MN while limiting the number of messages which have to cross the satellite system.

5.6.3. *Specific cases*

During our experiments, we were able to observe specific behavior in certain situations, especially at the level of the RRT procedure connected to Mobile IPv6. Therefore, in the case of movement 3, only the CoTi/CoT messages were exchanged with the CN (thus reducing the interruption time by 1.2 s). Likewise, for movement 4, where only the exchange of HoTi/HoT is usually necessary, no messages in the RRT procedure were exchanged (thus reducing the interruption time by 0.6 s).

Johnson *et al.* [JOH 04] explain this specific behavior indicating that, in the case where an MN moves quickly and regularly between different networks, it can occasionally reuse a keygen which is still valid. This is particularly efficient for movement 3, for instance since no message in the RRT procedure is exchanged via the satellite system.

However, we cannot consider these specific cases as opportunities for improvements since this would disrupt the RRT mechanisms as a whole.

5.6.4. *Problems related to overheads*

Another advantage of SIP mobility is that it does not add any overhead to the packets exchanged between the MN and CN when the MN is in a visited network, unlike the solutions based on Mobile IPv6. Indeed, when the MN communicates directly with the CN from a host network, Mobile IPv6 adds an extra 24-byte field to the header of each IPv6 packet in order to indicate the HoA of the MN and thus make the mobility of the MN transparent: the field "Destination Option Header" from the MN to the CN and the field "Routing Header" from the CN to the MN.

The IPv6 header, therefore, increases from 40 to 64 bytes, which can be particularly detrimental for VoIP communication which generally uses small user datagram protocol (UDP) packets. If we take the example of a GSM codec which sends a packet every 20 ms whose payload is 33 bytes (giving a useful flow rate of 13.2 kbps), the flow rate at the IPv6 level (taking account of the real-time protocol (RTP), UDP and IPv6 headers) increases from 37.2 to 46.8 kbps, constituting a rise of around 26% (three audio streams using the Mobile IPv6 header are equivalent to four streams without the Mobile IPv6 header).

This overhead grows further during the bidirectional tunnel phase since this involves an IPv6/IPv6 encapsulation. If we take the example of the GSM codec, the IP flow rate rises to 53.2 kbps and the same result is given for all the phases where a tunnel is implemented such as FMIPv6, PMIPv6 and HMIPv6 (for the latter, in the event of MAP domain change, there is even a phase of IPv6/IPv6/IPv6 double encapsulation). This, therefore, means a greater consumption of resources which is detrimental, especially for the WiFi and satellite connection.

This problem of overheads confirms the fact that in a satellite system where the resources are limited, the use of RO is essential not only in terms of delay in the transmission of packets but also in terms of bandwidth consumption. However, in any case, a solution based on SIP is less costly in terms of the use of resources for an equivalent quantity of useful data since it does not add any extra overhead.

5.7. Conclusion

These various comparisons show that the solutions based on Mobile IPv6, although they have the unquestionable advantage of managing mobility in a transparent way for all applications, are generally less efficient than the solutions based on SIP for mobility management for multimedia applications. Therefore, the experiments carried out on the satellite simulation platform concur with the theoretical evaluations, conclusions and recommendations made in section 5.4.2. Indeed, the efficiency of solutions based on Mobile IPv6 in terms of interruption time is reduced to specific cases of mobility (mainly returning to the home network and micromobility, which were not studied in the experiments), while SIP mobility is more consistent in the general management of the different movement types.

This becomes even clearer for networks with limited resources, as it is the case for a satellite system, since solutions based on Mobile IPv6 add an overhead which can be significant as soon as the MN is no longer in the home network.

6

The Transport Layer in Hybrid Networks

6.1. Introduction

As discussed in the previous chapters, satellite networks have all the technical characteristics enabling them to be an access technology for future hybrid or integrated network architectures, supporting multiservices (type 3G+ and 4G), with everything which that entails in terms of requirements linked to new uses: general and transparent access to services, with mobility, everywhere, all the time ("ubiquitous").

Due to their wide coverage capacity (user access) or broadcasting efficiency (point-multipoint), satellites can be very attractive as a complement to cellular terrestrial solutions. However, they come with a number of restrictions, making their use very specific and less transparent in terms of next-generation network (NGN) architectures, built around a technological Internet Protocol (IP) core.

This is especially the case for the "high-level" protocols of user terminals, whether they are personal users, smartphones or mobile tablets. Indeed, although it is possible to adapt communication protocols in the network, via special architectures deployed by the service provider, it is much more difficult on the user terminal which must make do with the standard configuration, suitable for the "general" Internet.

The first layer of the open systems interconnection (OSI) model which encounters this problem is the transport layer, which is the first "end-to-end" layer, since the software necessary for managing it is only deployed in the entities at the extremities: clients and servers. In the range of Internet transport protocols, there are two main protocols: transmission control protocol (TCP), mainly for reliable communications, and user datagram protocol (UDP) for the rest. TCP is now the most extensively used transport protocol. A study dating from 2009 [LEE 09] shows that the evolution in the use of TCP in the Internet has been relatively stable over the last decade. Over 90% of bytes transferred over the Internet have used TCP. This is not quite the same in terms of streams, because a little less than 30% of Internet streams used UDP. TCP is mainly used for the transfer of large volumes of data and UDP for small connections.

Since the advent of the Internet, the TCP/IP model (technology from local networks) has appeared to be less optimal, in terms of protocol performance, in certain wide area network (WAN) networks and especially in restricted environments such as satellite networks. Therefore, a number of solutions have been developed to enable the correct operation of TCP in satellite networks. In 1997, the Internet Engineering Task Force (IETF) created a working group called "TCP Over Satellite" which produced two documents in 1999 summarizing the standard solutions for optimizing TCP for satellite networks [ALL 99] and specifying the possibilities for future improvements [ALL 00]. Likewise, the Consultative Committee on Space Data Systems (CCSDS) deemed it necessary to participate in the standardization of communication protocols in order to cover the specific needs in terms of data transmission during space missions. A first version of a protocol stack named Space Communications Protocol Specifications [SCP 06] was produced in 1999, and was revised in the 2006 version.

Thus, the beginnings of the TCP protocol especially for satellite networks emerged, maximizing TCP performance in this environment, but which encountered problems in their deployment over user terminals. The solution identified, which is still being used, was to fit equipment between the satellite network and the terminals in order to transform the operation of the TCP protocol used into a satellite-compatible version. This equipment, known as performance enhancing proxies (PEPs), whose principle was standardized by IETF [BOR 01] in 2001, is distributed throughout satellite networks, offering both advanced services and web caching.

The solutions of the transport level used today, whether derived from TCP, UDP or based on the use of PEPs, are more and more limiting for satellite systems or are on the other hand too aggressive for the network. Above all, the improvement in terms of performance that current PEPs offer is not easily compatible with a number of deployment scenarios, especially in the military or aeronautical service sectors, due to security or mobility restrictions which could become rapidly more widespread in the traditional civil communication sector. Indeed, protocol optimization solutions (typically known as PEPs), commonly used for satellite connections, are not "transparent" enough for satellite/terrestrial hybrid networks. Therefore, they constitute a real obstacle to the "convergence" approach, targeting the integration of the satellite in terrestrial networks for mobility services.

Therefore, this chapter addresses the problems related to the transport layer, and especially TCP, in satellite/terrestrial hybrid networks. Section 6.2 will examine further the principle of PEPs and why they pose a problem in this context. Sections 6.3 and 6.4 looks at the evolution of TCP which has significantly evolved in the last few years and will show how it is possible to develop alternatives to the traditional solutions usually put forward in the context of satellite connections. Finally, section 6.5, the specific case of hybrid networks will be examined.

6.2. Performance enhancing proxies

This section presents two adaptations of TCP to satellite environments: SCPS transport protocol (SCPS-TP) (detailed in the reference document [SCP 06]) and I-PEP [ETS 09d]. Among the protocols defined by the CCSDS (listed below as reminder) is the SCPS-TP. This is the only protocol to have enjoyed commercial success because it was used as a basis for PEP developments in order to improve TCP performance over geostationary satellite connections.

More specifically, the SatLabs consortium supported by ETSI adopted the SCPS-TP protocol as a basis for the I-PEP specification. I-PEP was put forward to help the implementation of interoperable PEPs between spatial equipment (mainly, digital video broadcasting – return channel via satellite (DVB-RCS) solutions). This specification is itself based on the CCSDS specification of SCPS-TP.

6.2.1. *Space communications protocol specifications*

6.2.1.1. *SCPS-TP*

As a reminder, the SCPS protocol stack specified by the CCSDS consists of:

– *SCPS-FP*: a set of file transfert protocol (FTP) extensions to make it more efficient and add more advanced functionalities (record update and file integrity control);

– *SCPS-TP*: a set of TCP-modifying options on the sender side in order to improve TCP efficiency in restricted environment (long delay and high level of binary error or asymmetries). SCPS-TP options are TCP options registered with internet assigned numbers authority (IANA) and therefore SCPS-TP is compatible with other TCP implementations;

– *SCPS-SP*: a security protocol comparable with IPSec;

– *SCPS-NP*: a "bit efficient" network layer equivalent to IP but not interoperable with it.

SCPS-TP offers a number of transport services (reliable or not, connected or not and with or without acknowledgment), as an extension of the current basic Internet protocols, such as TCP, T/TCP and UDP which are:

– a "*fully reliable*" service which guarantees correct reception of all data, without errors and in order;

– a "*partially reliable*" service which guarantees that the data sent were received without error and in order, but does not guarantee the completeness of the transmission;

– a *non-acknowledged* service which simply guarantees reception without errors (order and completeness are not guaranteed).

Modifications are focused on satellite communication networks with restricted resources. They, therefore, satisfy the specific demands of spatial connections: significant/variable propagation delays (asymmetrical connection), reduced bandwidth (allocated on demand) and random/sporadic dropping of packets.

Here, we have listed the extensions proposed by SCPS-TP to improve TCP:

– *TCP for transactions* [BRA 94a]: simplifies the "handshake" upon the initialization of the TCP connection and offers a "reliable datagram" mode for telemetry/remote control (TM/TC)-type traffic (useful for very long distance connections);

– *window scaling* [BRA 92]: for communication environments requiring more than 65 k bytes of data in transit (at any given time);

– *round trip time measurement* [BRA 92]: for connections with high bit error rates (BERs), variable delays or a significant quantity of data in transit at any given time;

– protection against "*Wrapped Sequence Numbers*" [BRA 92]: responds to environments with high propagation delays or very high bandwidth;

– *selective negative acknowledgment* (adapted from [FOX 89]): designed for connections with high drop rates (high BER);

– *selective acknowledgment* [MAH 96]: also for connections with high error rates (high BER);

– *record boundary indication*: possibility of marking and transporting end-of-record indications from end-to-end in a reliable way;

– *"best effort" communication*: a paradigm enabling an application to retrieve data received, without errors and in sequence, but possibly incomplete;

– *header compression* (adapted from [JAC 90]): for environments with low bandwidth;

– *minimal congestion control* (low drop rate) or no congestion control;

– *explicit congestion notification* (ECN) [RAM 01]: offered as a way to improve performance for an ECN-compatible network;

– *specific retransmission strategy* for spatial environment enabling a better handling of problems related to corrupt data, dropped connections and congestion.

6.2.1.2. *Services made reliable with SCPS-TP*

This operating mode is entirely based on the TCP Internet standard as well as on the request for comments (RFCs) which are related to it, with a number of extensions and options specified in section 3 of the document [SCP 06].

The TCP extensions affect:

1) the establishment of the connection: negotiating SCPS-TP options "supported capability";

2) transmission (management of "record boundaries" and the "best effort" transport service");

3) error management;

4) the control of streams (congestion and transmission);

5) timeouts;

6) dropping of the network connection;

7) selective negative acknowledgment (SNACK);

8) header compression;

9) multiple transmissions (used as a "forward error correction" technique as an alternative to flow/congestion control mechanisms in a closed loop such as Van-Jacobson or Vegas).

Without detailing all the SCPS-TP options, the main options with a significant influence on the performance of this protocol are linked to techniques of retransmission, and stream and error control. In particular, the algorithm for calculating retransmission timers is replaced by a version integrating shaping based on round trip time (RTT) and its variance. A mode with or without congestion control is possible. The first (without) deactivates the "slow start" and avoids TCP congestion. The second (with) offers the use of standard congestion control (Van Jacobson and exponential back-off) or the use of Vegas.

If the network enables the root cause of the dropped packet to be ascertained (actual congestion or loss/alteration due to transmission), then it is advised to apply the "exponential back-off" mechanism only for the case of loss/alteration. This test of actual congestion can be retrieved from a management information base (MIB), and/or a cross-layering mechanism. This may also be the case, via ECN [RAM 01], if ECN has been implemented.

The specification of SCPS-TP also provides some recommendations regarding the measures to take in the event of a break in the connection, when the TCP emitter is notified (by any means: cross-layer, or via an MIB request).

The implementations which need to comply with the recommendation given in the specification, and establish a compliance sheet, detailing the choices adopted for each of the options open in the SCPS-TP specification.

6.2.2. *I-PEP*

Like SCPS-TP, I-PEP is an implementation specification for DVB-RCS satellite communication. I-PEP extends SCPS-TP (and is therefore compatible with it).

6.2.2.1. *The IPEP objective*

I-PEP specifies PEP interoperable behavior in order to provide the facility of satellite equipment integration via various providers such as the one Figure 6.1 shows. To do this, it is necessary to standardize the behavior of the PEP satellite interface in order to authorize the mixing of PEP software of various vendors.

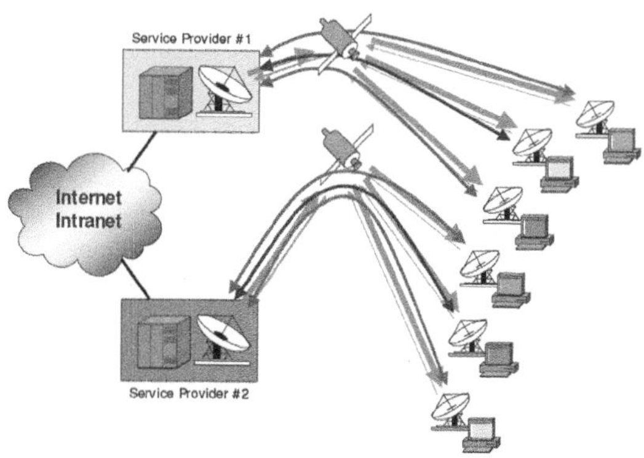

Figure 6.1. *General view of I-PEP [ETS 09d]*

Functionally, I-PEP focuses on the air interface (satellite connection). I-PEP depends on a "split" mode PEP architecture, transparent for terminal applications (Figure 6.2). The notions of PEP client and PEP server, commonly used to distinguish PEP entities, are relatively artificial since the I-PEP protocol is symmetrical.

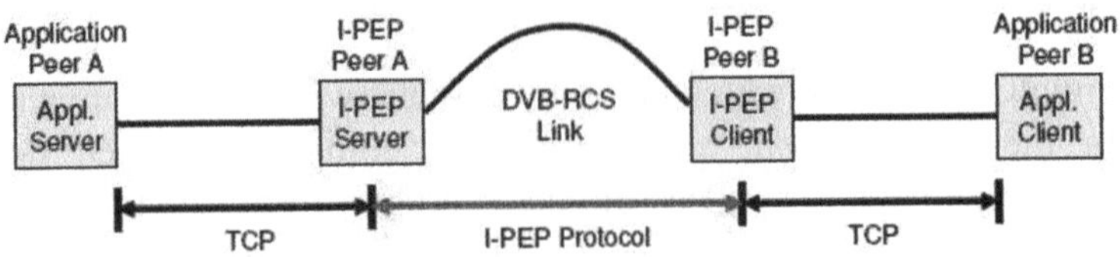

Figure 6.2. *Basic I-PEP components [ETS 09d]*

Figure 6.3 shows three non-exhaustive examples of the integration of the I-PEP protocol: (a) several application clients served by the same I-PEP client, (b) a configuration in which the client application (user equipment) is associated with a single I-PEP client, in relation to several I-PEP servers colocated over the satellite hub and (c) an I-PEP client for a distant site (DVB-RCS integrated terminal) linked to an I-PEP server which is not colocated over the satellite hub.

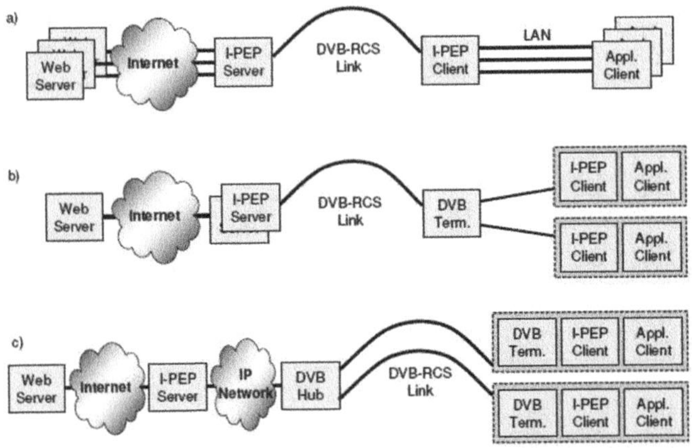

Figure 6.3. *I-PEP protocol integration scenarios [ETS 09d]*

The I-PEP transport protocol is used to transport packets in a reliable way between two PEP entities (client/server). To avoid defining a new protocol from scratch and ensure the interoperability with non-PEP entities, TCP is used as a base.

As has been demonstrated many times, TCP is not particularly suitable for communications routed through networks whose bandwidth delay product (BDP) is high or which are subject to transmission errors (not caused by "real" congestion). By definition, satellite connections are in the category of high BDP, and depending on the conditions may also fall into the second category.

6.2.2.2. *Overview of the protocol*

Above all, it is a transport layer (in compliance with TCP), and optionally a session layer.

I-PEP functions in "split" mode: the transparent or proxy modes are possible. The I-PEP connection between the client and the server functions in two modes: "hop-by-hop" or "end-to-end".

I-PEP provides a communication channel between a client and a server enabling the optimized exchange of data between an application client (TCP) and a server, typically used to access the Internet. I-PEP transport offers the following services:

– reliable communication: ordered (sequenced), protected against error and with stream control;

– minimization of the bandwidth used, limiting the protocol overhead;

– optimizing communication performance, by reducing the number of satellite two-way satellite links and defining a transfer-rate and congestion control suitable for satellite connections.

Moreover, a session protocol is added to the transport layer, whose minimum requirement is:

– determining the location of the server;

– and negotiating the minimum characteristics for exchanges between I-PEP entities, the network characteristics to take into account by I-PEP (e.g. maximum return transfer rate, minimum bandwidth allocated and hypertext transfer protocol (HTTP) acceleration supported).

Optionally, the session protocol can propose:

– authentication of client equipment (and thus the user);

– security of communication (encryption over satellite connection).

By extension, the specification provides recommendations for application improvements such as HTTP prefetching, or specific application protocols.

This enables proprietary extensions to the I-PEP protocol, enabling added services (functional and performance). There is no application programming interface (API) for the I-PEP service, since the interface remains fully compatible with the TCP access.

Session management

– localization of server: to ascertain the IP address of the I-PEP server(s): by announcement (signaling) over the satellite connection, by domain name server (DNS) request, or by configuration of the client;

– session establishment: a session is associated with a set of transport communications, enabling the sharing of a communication context for them. The minimum enables the negotiation of transport options (options supported by the implementation of the I-PEP protocol, as well as the identification of the provider and the protocol version);

– the session can also be used to authenticate the client, negotiate a shared secret, as well as other services such as billing;

– each session is referenced by a unique identifier;

– termination of the session: by the client of server via an explicit message or when there is no longer transport communication (implicit);

– session maintenance: if necessary, there is an exchange of control messages in order to supervise the client/server communication or to update context parameters.

6.2.3. *Issues related to PEPs*

PEPs are a reality in satellite offers. They are mainly used in satellite network architectures, both for the performance optimization they provide and also for the application acceleration which they can offer (HTTP and DNS). PEPs have, therefore, significantly exceeded their initial objectives, offering more and better solutions than those possible with the initial restrictions of the operation of TCP over satellite.

However, PEPs still come with a number of restrictions which make their use problematic in the architectures of NGNs. The following sections briefly show the problems which have emerged related to the security and mobility of services.

6.2.3.1. *Impact on security*

PEPs need to analyze the headers of TCP segments and IP packets exchanged between border entities in order to initiate where necessary a suitable connection for the network to cross. The virtual private network (VPN) offers, whether via satellite or not, generally use IPsec tunnels to secure communications, thus masking the content of IP packets and in particular the source and destination of data. Therefore, it is impossible to implement a PEP via a tunnel. Nevertheless, it is possible to install PEPs at the ends of the IPsec tunnel, but this requires more specific configurations, security which is not necessarily end-to-end and the necessity to know specific connections crossed by the tunnel. In the event where the tunnel is open from the user terminal (preferable from an information security point of view), the satellite terminal (ST) can no longer act on the content of packets, as it is necessary, for example, for the classification in quality of service architectures, or processing by a PEP.

6.2.3.2. *Impact on mobility*

The implementation of mobility architecture, such as mobile IP seen previously, also poses a number of complex problems for PEPs. The biggest obstacle is caused by hybrid movement. This is in the event where a satellite network requiring acceleration by PEP to a network where this acceleration is no longer necessary, and even restrictive. Indeed, TCP connections managed and accelerated by the PEP should be able to continue without PEP, without transport disconnection and via another pathway. There are three possible scenarios to enable this [DUB 10]:

– mobility is managed by the lower layers of the communication system making the transport layer transparent. The PEP should, therefore, be used over the terrestrial network, along with the problems this creates (aggressiveness);

– mobility is managed by the higher layers. In this case, TCP connections over satellite that use a PEP should be directly sent to the receiver when a terrestrial network is used. The conversion of a sequence of two TCP connections into a single running end-to-end TCP connection is then needed. This is not currently possible for standard TCP stacks;

– mobility is managed by higher layers, but we accept that we must rely on an intermediary element in the new network. This new PEP, which will generate superfluous processing, may be adapted to the new network, even if this is not necessary over a terrestrial network. Therefore, a synchronization of PEPs must be carried out via an exchange of contexts. During this transfer, TCP connection must be paused so as not to disturb communication. PEPs must be "compatible" enough to accept this complex situation. There is currently no standard defining such a context transfer.

Therefore, PEP use necessary for the optimization of the TCP protocol over satellite networks is a significant obstacle to the integration of satellite networks into next-generation terrestrial networks. Section 6.3 will examine the recent TCP progress made which could offer an optimal solution to the problem.

6.3. TCP evolutions

In its 40 years of existence, the TCP protocol has changed greatly in order to adapt itself to the evolution of networks and it would be impossible here to give a detailed history of all these changes. However, we will briefly discuss the main modifications which it has undergone, especially in terms of congestion control. Created in 1973 by ARPANET, and known under the name Transmission Control Program (RFC 675), it laid the foundation of what would become the TCP protocol. The programming interface, data structures and connection management mechanisms are defined in this document. At the end of the 1980s, the RFC 1122 specified that an Internet transport protocol must implement congestion control, and above all "slow start" and "congestion avoidance" mechanisms based on a variable sized window. This led to the production of TCP Tahoe, the first version of TCP using congestion control as it is known today. The "slow start" phase looks to exponentially increase the size of the emitter's congestion window until it causes dropping due to congestion in the network. The "congestion avoidance" phase which follows causes a linear increase in the size of the window as long as there is no dropping.

The main problem with TCP Tahoe lies in its drop detection, based on timers, which are often too long in comparison with a shortening network delay (twisted cables, fiber optics, etc.) The "fast recovery" and "fast retransmit" mechanisms were added 2 years later to enable faster and less detrimental detection and correction for the congestion window, giving rise to the TCP Reno version. In 1994, a new congestion control version was released, under the name TCP Vegas, which detected congestion by analyzing drops and the evolution of the network crossing delay. The delay is measured by TCP segments and their acknowledgments. An increase in this delay is interpreted as in increase in network buffering, leading to a congestion risk. Despite the implementations available and a very good transfer rate adaptation, TCP Vegas was never widely used.

TCP new Reno (RFC 3782) improves the "fast recovery" mechanism by carrying out the re-emission of a segment at each duplicated acknowledgment. This TCP version increased the emission rate compared to Reno, especially over connection with a high drop rate. TCP New Reno is currently the most widespread version of TCP, but could soon be replaced by the new versions discussed in section 6.3.3.

6.3.1. *TCP adaptations to the satellite environment*

Many TCP versions have been proposed in order to adapt it to suit a specific medium or environment, generally wireless or satellite. It would be unwise to review all of these different versions, therefore, only the most significant and recent ones regarding the satellite environment have been included.

These specialized versions are not widely deployed (especially on the server side) and therefore cannot be separated from the PEP solution:

– *Noordwijk* [KRI 08]: this protocol, developed by European Space Agency (ESA), was designed and implemented in order to optimize the transfer of the smallest packets (e.g. HTTP) over a satellite connection, while ensuring good performance for the transfer of large packets (e.g. FTP). Its target context is a communication environment adopting the DVB-RCS standard with demand assignment multiple access (DAMA) between two PEP entities (see the description of PEPs in section 2.2.4).

– *TCP FIT* [WEN 10]: TCP FIT is based on the idea of using several TCP connections, as certain applications do, in order to achieve the best transfer rate. This version carries out an estimate of output packets and the average size of the window at the last period in order to decide whether to increase or decrease the number of virtual connections. The congestion window then changes as a result. Experiments show a TCP Friendly behavior from TCP FIT and an improvement in the transfer rate of up to double compared to Cubic in simulated long-term evolution (LTE) networks;

– *Hybla* [FIR 04]: TCP Hybla is an improvement of TCP which fights against performance deterioration caused by the long RTTs typical to satellite networks. In the context of a heterogeneous network containing a high latency segment (e.g. satellite), Hybla aims to eliminate the detrimental effect of high latency segments compared to segments with small RTTs. It is made up of procedures including: the Hoe algorithm [FIR 04] for estimating the capacity of the channel, timestamps, adoption of the selective acknowledgment (SACK) policy and packet spacing techniques. TCP Hybla only entails a modification on the emitter side. This enables full compatibility with standard receivers.

6.3.2. *Options and mechanisms for TCP improvements*

This section summarizes the main mechanisms and options for improving TCP or adapting it to certain types of networks such as WiFi and satellite networks. Some of them are generic (e.g. TCP options), while others are specific and change the TCP behavior, or alter TCP operation under certain conditions or in certain contexts.

6.3.2.1. *TCP options*

TCP options can be useful since they do not modify the semantics of the protocol and can be activated/deactivated depending on their use. The ones presented below are listed by IETF and have an IANA identifier:

– *TCP sack*: TCP acknowledgments are cumulative. Only the value of the first non-acknowledged segment is returned by the receiver. The SACK option tells the emitter which segment has been dropped. SACK [MAH 96] notifies the emitter of up to four blocks of segments received. If some of them are dropped or out of order, this is indicated in the option field. This is even more useful given that wireless and satellite connections are often subject to correlated drops. An improvement of the fast retransmit algorithm is based on this option [ALL 03] and some TCP versions use it to increase efficiency. This option is negotiated during connection initialization;

– *TCP timestamp*: the RTT measure is important in determining the retransmission timeout (RTO). However, in TCP, a single RTT value is calculated per window which makes it potentially incorrect. The timestamps option [BRA 92] contains two fields: the time value of the emitter and – for acknowledgments only – the initial time value of the packet which has been acknowledged. This functionality can be used for a number of other reasons, notably, for detecting reorganization or for identifying a received segment which has been acknowledged. This option must appear in a synchronize (SYN) segment in order to be activated;

– *TCP window scale*: this option [BRA 92] helps increase the size of the TCP receiving window over its maximum value. Each host can use a different scale factor for its own window. The value is sent at the moment of connection initialization. By defining the field at 0, a host can give the user this option without affecting its own window. This option is very useful in satellite networks since it means that the window does not have to limit the emission rate linked to the product of the window size per the connection delay;

– *TCP user timeout*: in standard TCP versions, the connection times out due to non-acknowledged segments. This option [EGG 09] announces the timeout in order to share the time policy. A longer timeout means longer periods of silence in the event of mobility, for instance, while a short timeout means that contact needs to be maintained in order to preserve the connection.

6.3.2.2. *TCP optimizations*

Unlike the TCP options above, this TCP optimization is not standard and can, therefore, not be easily implemented for a connection unless there is an adapted terminal at the emitter and receiver ends of the network. These mechanisms are, therefore, possible TCP improvement solutions behind a PEP:

– *initial window*: the slow start TCP phase is necessary for achieving an acceptable transfer rate in the network and it also adds to the latency right at the start of the connection. In 2002, the initial size of the congestion window was revised enabling an increase of up to four segments. Today, websites are becoming increasingly complex and Google requested that IETF increase the initial window to 10, since it had observed that it was sufficient for downloading 90% of web pages in a single RTT.

Consequences in terms of congestion management appear to be negligible in most environments, but may be problematic if the bandwidth is limited. A smaller window can be used if this is the case. If this change is implemented in the TCP stack, it significantly improves the HTTP behavior in long delay networks such as satellite networks;

– *quick-start*: congestion algorithms take time to adapt to the correct Cwnd, which can lead to problems over a variable support like a wireless network and above all in case of vertical mobility. Quick-start [ALL 07] is a cross layer mechanism between IP and TCP enabling an explicit transfer rate requirement for the whole access pathway;

– *early retransmit*: [AVR 10] this triggers the fast retransmit, in certain conditions, especially when few segments are exchanged, in order to reduce the number of "duplicate acknowledgement (ACK)" necessary for a fast transmission (normally three). This enables the use of fast retransmit to retrieve dropped segments which, otherwise, require a long retransmission delay with timeout. In other words, connections recover from a drop faster, thus improving latency;

– *limited transmit:* in the environments where the RTT is high, the time lost can significantly delay a connection. When a reorganization occurs, TCP waits for three duplicate ACKs before activating the fast retransmit, but the arrival of the first two duplicate ACKs indicates that the number of packets in circulation has fallen (acknowledged data). It is possible for the TCP protocol to send two new segments if authorized by the pair's advertised windows (Awnd).

6.3.3. *New TCP versions*

TCP is constantly evolving to adapt to new communication paradigms offered by the new supports (wireless, etc.) and new uses (high speed, video, etc.). However, the use of a TCP version depends above all on its ability of being rapidly deployed in the hosts at the extremity of the network. Today, this is only possible through diffusion in new systems or possibly specific PEPs. Indeed, despite burgeoning research activity, the TCP protocol can only evolve in an infrequent piecemeal manner.

Nevertheless, the situation has changed a great deal in recent years with the appearance of two new modes of TCP congestion control, promoted on one side by the Linux community (CUBIC) and on the other side, by Windows (Compound). Due to their worldwide distribution and frequent operating system (OS) updates, these versions are now widely used:

– *Cubic* [RHE 08]: in this case, the congestion window changes according to a cubic formula involving the time since the last congestion event. Cubic does not wait for the ACKs to increase the size of the congestion window, since Cubic's Cwnd only depends on the previous congestion event. Cubic is the default version in Linux systems with a kernel above or equal to 2.6.19.

The emission window is controlled by the equation:

$$W_{cubic} = C(t-K)^3 + W_{max} \quad [6.1]$$

(TCP Cubic control of the congestion window)

where C is a CUBIC parameter, t is the delay since the last reduction of the window and K is the delay for increasing W to Wmax when there has been no drops;

– *Compound TCP* [ZHA 06a, ZHA 06b]: CTCP was designed to rapidly adapt to the bandwidth available while remaining TCP friendly. Its main particularity in its Cwnd management is that it is based both on drops and delays. CTCP, therefore, manages a double congestion window: as is the case for TCP Reno and another based on the delay known as delay congestion window (Dwnd). This is only used during a congestion avoidance phase. For example, in a satellite network, the significant propagation delay can distort the retransmission timers and lead to drops. The congestion window based on delay enables an estimation of the end-to-end delay and takes it into consideration. CTCP is currently implemented over the operating systems Windows Vista, Seven and Server 2008. A patch is also available for the systems Windows Server 2003 and Windows XP 64 bits. Congestion control based on delay has the distinct advantage of anticipating congestion by measuring the evolution in the delay, and therefore limiting the drops which can be very detrimental to satellite networks.

$$win = \min(cwnd + dwnd, Awnd) \quad [6.2]$$

(Equation of the evolution of the CTCP congestion window)

where cwnd and the congestion window are based on drops (like TCP Reno), dwnd is based on delay window and Awnd is based on the reception window. Dwnd is calculated as shown below:

$$dwnd(t+1)=\begin{cases} dwnd(t)+(\alpha\cdot win(t)^{k}-1)^{+}, \text{ if } diff<\gamma \\ \left(dwnd(t)-\zeta\cdot diff\right)^{+}, \text{ if } diff\ \geq\gamma \\ \left(win(t)\cdot(1-\beta)-cwnd/2\right)^{+}, \text{ if a loss is detected} \end{cases} \quad [6.3]$$

(Equation of the evolution of the CTCP congestion window based on delays)

where diff is the difference between the estimated transfer rate and the rate measures, multiplied by RTT. This corresponds to the number of packets inserted into the network, but which could not be received and are therefore in transit in the network. γ is a trigger threshold and the values of α, β and ζ are constants.

6.3.4. *Characteristics of the satellite connection*

Geostationary orbit (GEO) satellites are the most widely used due to their fixed position in relation to the Earth. Permanent coverage of part of the planet can, therefore, be ensured. This is why GEOs are mainly used for telecommunications, despite their high cost and propagation delay.

The satellite communication method is characterized by a number of specific aspects which are examined below.

6.3.4.1. *End-to-end delay*

The raw propagation delay in a GEO satellite network is around 250 ms, which is a rather large delay given the transmission and propagation characteristics. By adding the emission time, access time, pathway time in the underlying networks and processing time (congestion control, retransmission, coding/decoding, encryption/decryption, etc.), the end-to-end delay can reach values much higher than the raw figures. Figure 6.4 shows an evaluation of an application delay (ping echo request/reply) over a real satellite (DVB-S2/RCS platform in Ka band OURSES [OUR 10]). The measure shows the evolution of the RTT between a client behind an ST and a server behind the gateway. An average request/reply delay of 660 ms can be observed.

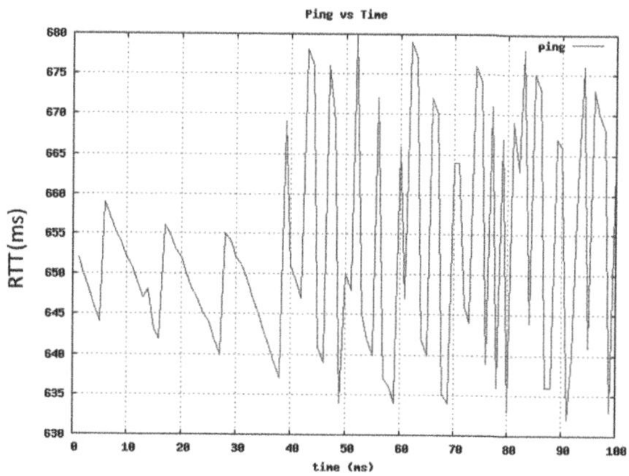

Figure 6.4. *Request/reply delay of a ping over a real satellite connection – OURSES platform*

Bandwidth limitation

The cost of the bandwidth in the satellite network remains high. This reflects the low transfer rates offered in the current commercial systems, especially over the DVB-RCS return channel. However, with the application of little-used band (Ka), the improvement in adaptive coding and modulation (ACM) and the multispot, it is possible to offer a high-speed service.

6.3.4.2. *Signal to noise ration (SNR), signal reduction and the impact of drops*

Errors due to the transmission channel are common over a satellite connection. They are caused by the air interface which varies because of changes in the radio signal.

Nevertheless, some of these particularities are also present with terrestrial wireless technology, especially errors. They have been reduced due to improvements which have been made to the radio interface, such as coding algorithms and ACM techniques. The DVB-S2 has already been proven and provides the satellite with a channel which has a very low error rate.

6.3.4.3. *Asymmetry in the transfer rate between the forward and return channels*

While providing cost-effective and efficient services, the transfer rate over the DVB-S2 forward channel is significantly higher than that on the

DVB-RCS return channel. This is due to the nature of satellite radio transmission and the scales of the systems.

6.3.4.4. *Jitter variation*

To compensate for the variation in the channel (fading or deterioration of signal), the satellite implements mechanisms for control, coding and adaptive modulation (ACM) leading to a jitter variation for the use.

6.3.4.5. *Access framework in bursts*

The access channel is not continuous, but is based on a slot allocation. This leads to data transmission in bursts, occasionally creating high levels of jitter.

6.3.4.6. *Bandwidth on demand*

The allocation of bandwidth is often centralized. It uses mechanisms such as DAMA. This leads to a significant extra delay before obtaining the requested bandwidth. Capacity and delay may change abruptly over a given period. Not all types of satellite communication systems include bandwidth on demand – some of them only have static allocation.

6.3.5. *Impact on the transport layer*

All these characteristics of the DVB-S2/RCS satellite system have an impact on the behavior of the TCP transport layer. However, we will focus on the significant impact of the long delay and low transfer rate on TCP communications.

Above all, for TCP in satellite networks, errors are no longer the main problem, compared to the long delays and variable transfer rates.

Standard TCP algorithms such as slow start, congestion avoidance and the use of RTO timers are not designed to operate over satellite networks due to the long delay, but also the asymmetry and variation of the transfer rate.

In short, the main TCP problems in a satellite environment are:

– A long RTT leading to:

- longer slow start phases, involving a dropping of resources and bandwidth,

- a longer error detection process, because, for instance, the TCP retransmission timer (RTO) is proportional to the RTT.

The latency introduced by the on demand bandwidth mechanisms may increase the RTT. This involves a slow evolution of the size of the congestion windows during the slow start phase with a possible RTO timeout;

– A large BDP creates a large transmission window for each TCP connection. Inevitably, this increases the likelihood of packets being dropped and may trigger TCP congestion control mechanisms. In most cases in satellite networks, the TCP congestion control mechanisms interpret drops as congestion. This has a direct impact on the user since it leads to a significant reduction in the transfer rate. Large RTTs have a negative effect on TCP behavior if it is not taken into account in these mechanisms.

The BDP specifies the quantity of data that the sender needs to transmit at any time in order to fully use the bandwidth available. The BDP is the product of the connection rate and the RTT.

$$\text{BDP}=\text{bottleneck BW (bit/s)}*\text{ RTT (s)} \quad [6.4]$$

(Bandwidth delay product BDP)

– *The asymmetry of the connection* can lead to a low capacity of returning ACKs [PAD 02], thus causing a desynchronization of ACKs and limiting the transfer rate over the forward channel.

6.3.6. *Conclusion*

This section offered a brief presentation of the main problems in the transport layers in satellite networks and the main solutions available today to improve the situation. The next section will focus on the measured behavior of the most common TCP versions.

It is not an exhaustive study, but it attempts to show, for instance, the result of the problems outlined above. Therefore, we will use specific cases to compare to operation of TCP versions (Reno, New Reno, Hybla, Cubic and Compound) and will show the advantages to expect from recent TCP versions.

The measures were carried out on Linux and Windows operating systems connected to a satellite network simulation test bed developed by AIRBUS DS (SATEM).

6.4. TCP performance in a geostationary network

6.4.1. *Measurement and analysis methodology*

The following metrics were chosen for evaluating the performance of TCP:

– the evolution of sequence numbers: this metric represents the evolution of the packets sent and is useful for the monitoring of the efficiency and regularity of a TCP connection. It provides information on dropped segments, breaks in connections and the quantity of data sent. The comparison of sequence numbers is interesting because it enables an easy comparison of different TCP versions;

– the size of the *CWND* congestion window used (not available for CTCP on Windows 7);

– the end-to-end delay which provides information on the state of the buffers;

– the application transfer rate which defines the average quantity of data sent at a given moment t.

The measures taken in this section were carried out with a single TCP connection over an empty link. The objective is to accurately observe the behavior of the satellite TCP congestion control.

These experiments show the data emission capacity of a TCP connection through the evolution of the sequence numbers. The end-to-end delay is measured as it is perceived by the application. This delay is then composed by the propagation delay and the time spent in the emission buffers. The application transfer rate is also measured.

6.4.2. *The configuration of the system and measurements*

The satellite connection is simulated with a one way delay (OWD) of 250 ms and a bandwidth of 512 Kbps or 2 Mbps, according to the test. The error rate is left at 0 (no fading). The results are shown in the following figure.

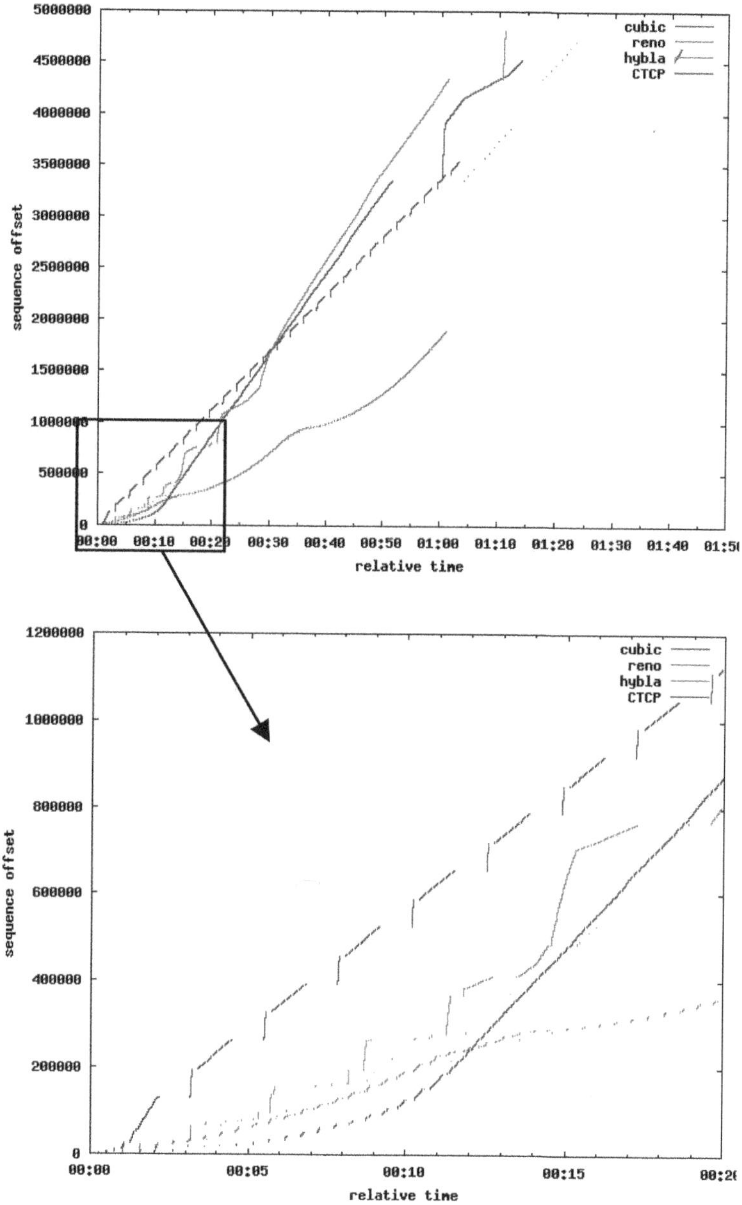
cubic
reno
hybla
CTCP
sequence offset
relative time
5000000
4500000
4000000
3500000
3000000
2500000
2000000
1500000
1000000
500000
0
00:00 00:10 00:20 00:30 00:40 00:50 01:00 01:10 01:20 01:30 01:40
cubic
reno
hybla
CTCP
sequence offset
relative time
1200000
1000000
800000
600000
400000
200000
0
00:00 00:05 00:10 00:15

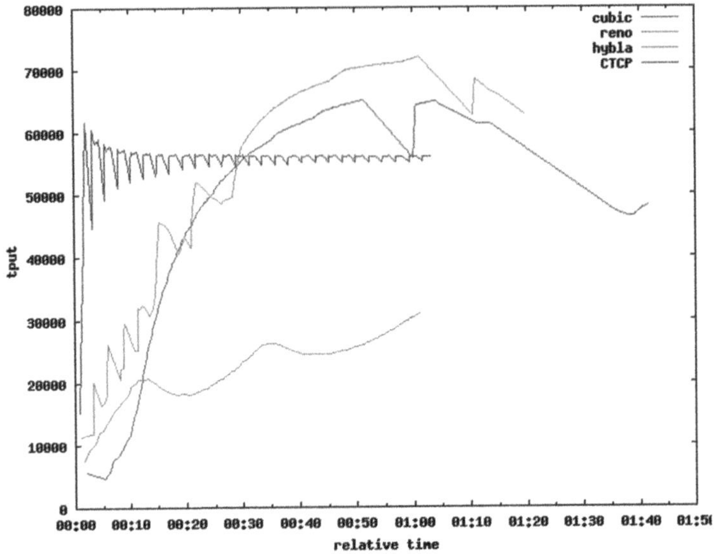

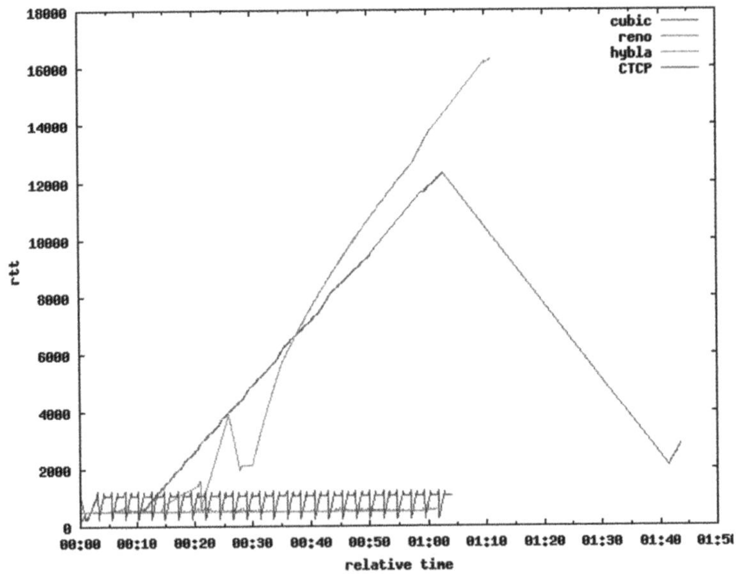

Figure 6.5. *Sequence number, transfer rate and RTT over a 512 Kbps connection. For a color version of the figure, see www.iste.co.uk/berthou/networks.zip*

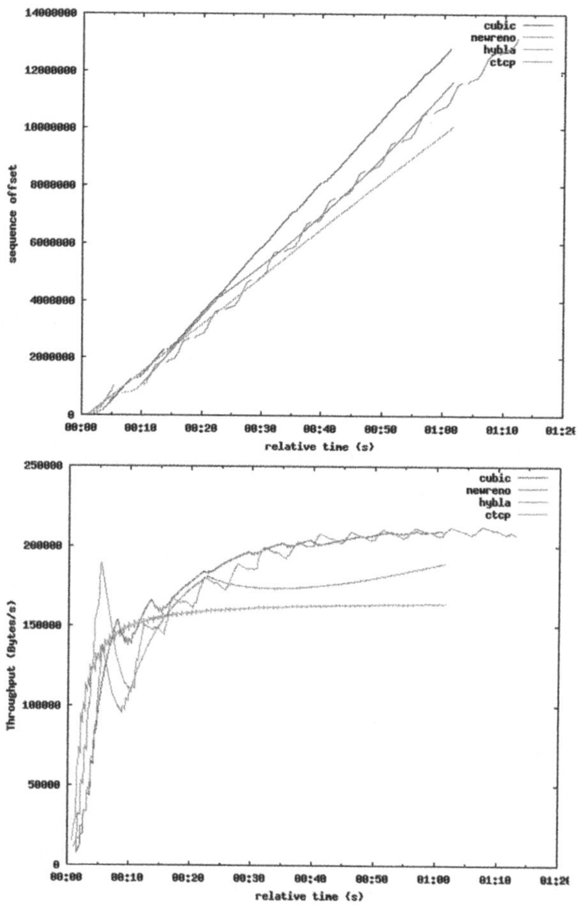

Figure 6.6. *Sequence number and the use of bandwidth over a 2 Mbps connection. For a color version of the figure, see www.iste.co.uk/berthou/networks.zip*

6.4.2.1. *Analysis*

In Figure 6.5, Cubic, Hybla and CTCP show better performance than New Reno. CTCP shows a transfer rate improvement of over 50% compared to New Reno.

Two periods stand out: one period from 0 to 30 s during which CTCP shows the best use of the transfer rate available and a second period, from 30 s until the end, in which the aggressive protocols such as Hybla and Cubic use the connection in the best way.

CTCP is the most stable version since its congestion control is based on the management of delays and drops. It performs better than the other versions due to a better estimation of the RTT.

The first few seconds of Hybla appear to be a little chaotic due to its aggressiveness. As a result, this triggers congestion control mechanisms (it is possible to observe some gaps in the progression of sequence numbers).

Over the first 30 s, the CTCP transfer rate is around 30% greater than for Hybla and Cubic. These two catch up with CTCP at the 30 s mark.

After t = 30 s, Hybla and Cubic show a transfer rate with a similar intensity as CTCP. This aggressive behavior leads to problems: 20 s later the transmission is interrupted for 10 s.

Consequently, the data stream does not stop at 60 s (test duration) due to the secondary effects of congestion and buffering (timeouts and retransmissions). However, CTCP demonstrates better behavior and does not encounter this type of problem.

In terms of bandwidth use and occupancy, CTCP quickly converges toward the correct transfer rate.

Hybla and Cubic take around 10 s longer to reach the same transfer rate, and then overtake it due to more aggressive algorithms. However, New Reno is not aggressive enough, leading to its poor occupancy of the channel.

In terms of retransmissions, Hybla retransmits over 2% of packets, compared to 3% for Cubic, 1% for New Reno and less than 0.1% for CTCP. New Reno's rate of retransmission is low but this is due to the total number of packets it sends, which is around one-third of the total packets emitted by the other versions. With this in mind, CTCP shows very good performance and has the best bandwidth use. This is important in the satellite context of limited resources.

Figure 6.6 shows the respective evolution of sequence numbers and the occupancy of bandwidth of the four different TCP versions with the same delay conditions (250 ms of OWD) and 2 Mbps of bandwidth. All streams have almost the same behavior with a slight advantage for Cubic. It uses the

2 Mbps better than the other versions, and reaches an approximate maximum value of 1.5 Mbps.

Hybla also shows the same behavior as Cubic but with more fluctuations. It adapts better to the long fat network (LFN) and its performances remain strong, with a transfer rate of 1.5 Mbps. Nevertheless, a number of drops can be observed in the initial packets until the stationary transfer rate.

New Reno also uses the bandwidth well and reaches the transfer rate of 1.3 Mbps, but encounters difficulties in the first 20 s. It is not designed for satellite networks but is still able to use the 2 Mbps better than the 512 Kbps.

Hybla which is designed for satellite networks and Cubic, due to its aggressiveness, overtake CTCP. However, it is important to indicate the instability shown by Hybla during the first 20 s.

CTCP reaches a transfer rate of 1.2 Mbps and has a transfer rate limitation when there is a high BDP. Having said this, a transfer rate of 1 Mbps for a single TCP connection (the problem does not arise when sharing 1 Mbps over several TCP connections) over a traditional commercial satellite system is an exception rather than a rule.

For retransmissions, Hybla retransmits over 7% of packets, compared to less than 1% for Cubic, 1% for New Reno and less than 0.1% for CTCP. Hybla encounters difficulties in LFNs. Cubic sends fewer retransmissions because it is well suited to LFNs [YAM 08]. However, CTCP shows very good performances in terms of retransmission of packets and has the best bandwidth use.

At first glance, CTCP appears to be the least efficient in this configuration, but according to the evolution of the windows and the transfer rate over the first few seconds, it is clear that CTCP reaches the stationary transfer rate faster. Moreover, CTCP is the most stable version, with fewer drops and provides an almost linear transfer rate evolution.

Management of the congestion window based on CTCP drops and delays gives it a faster adaptation, with a little lower transfer rate but a stable evolution.

6.4.2.2. *Heterogeneity of TCP versions*

Generally speaking, it is rare for TCP versions for a client and a server to be identical. Having shown the value of CTCP in a satellite context, Table 6.1 summarizes the experimental result showing a combination of the various client–server couples with different TCP versions.

In summary, the behavior of heterogeneous client/server combinations seems to be mainly influenced by the sender, with the exception of CTCP. Indeed, whether it is used on the sender or the receiver side, its use influences the overall behavior of the connection, which is close to a traditional CTCP homogeneous connection. New Reno benefits from the evolution of the receiver.

Receiver **Sender**	**New Reno**	**Hybla**	**Cubic**	**CTCP**
New Reno	x	Hybla +-	cubic +-	Ctcp +-
Hybla	Reno ++	x	cubic +-	Ctcp +-
Cubic	Cubic +-	Hybla --	x	Ctcp +-
CTCP	Ctcp +-	Ctcp +-	Ctcp +-	x
Legend : "--" worse , "+-" similar, "++" better				

Table 6.1. *Combination of different TCP versions (heterogeneous client/server)*

Table 6.1 indicates to which overall behavior of homogeneous connection the combination is most similar. When this behavior shows a slight deterioration or improvement, the notation – or ++ is used, respectively. +- indicates a behavior which is similar to the use of the protocol at each extremity. For instance, Hybla used on the receiver side and New Reno used on the sender side offer a similar behavior to a TCP connection where the sender and receiver are using Hybla.

These results are interesting given that the initial tests show the value of using CTCP in a satellite network. Indeed, a single modification of the servers' version of the TCP modifies the overall behavior of the TCP connections.

6.5. TCP in a hybrid context

6.5.1. *The impact of a hybrid network on the transport layer*

This section examines the problems faced by TCP in terms of handovers, in a mobility context and a heterogeneity of networks. We will particularly focus on hybrid handovers (horizontal or vertical) because in these scenarios the characteristic of the end-to-end pathway varies significantly when satellite/terrestrial handovers occur. Many articles have been published by the research community on TCP handovers, many regarding wireless or cellular terrestrial networks. Unfortunately, up to now, satellite networks have not been addressed. The main problems related to them are examined in the following section.

6.5.1.1. *The persistence of application based on TCP*

The first point for maintaining a connection of a TCP application during a handover is the TCP connection itself. The second point is to prevent the application from disconnecting.

Maintaining a TCP connection

TCP uses an IP address and a port number as a final socket identification point for the destination and source. If there is a change in one of these four parameters, the socket will become disconnected.

In the case of a vertical handover, the access network is switched from one access server to another, leading to an IP address reallocation (indeed, the subnetwork prefix needs changing in order to build an address which complies with the IP topology of the new network, and to authorize correct routing in the Internet network). TCP and IP were not designed to handle this type of problem.

There are two solutions for handling the change in IP address:

– *at the transport layer*: TCP extensions have been put forward to eliminate the dependence of the connection on the IP address. For instance, TCP-migrate adds two new options for the TCP protocol, and a new "state" (MIGRATE_WAIT) in the TCP state machine. This was not adopted since the impact it would have on the protocol would be too great. TCP-R (redirection) was another similar attempt, adding an extra layer to hide the broken connection of the transport layer;

– *at the network layer*: this is a family of protocols. Mobile IPv4 is the first extension introduced at the network level to solve this problem. Improvements or variations have been added in the continuity of MIPv4, MIPv6, fast handover mobile IP (FMIP), hierarchical mobile IP (HMIP), proxy mobile IP (PMIP), etc. The common principle of Mobile IPs is to hide the updated address from the higher layers.

Application timeout

Even though this connection is maintained at the TCP level, the application could not handle an interruption of the communication during a handover. The application may consider that the connection is lost (silence), which may lead to disconnection. A TCP option, the user timeout option (UTO) announces a timeout so that the application shares the timeout policy.

6.5.2. *Control of the adaptation of streams to the new network*

The maintenance of the IP address, and consequently the TCP connection, is not sufficient. There are changes in the connection characteristics such as the bandwidth and the propagation delay (and consequently the BDP) after a handover which poses serious problems to the management of TCP congestion.

TCP congestion control needs time to ascertain the capacity of the new network which may lead to an under-usage of the new connection (in the event of a handover toward the higher network with more capacity) or drops in the connection (in the event of a handover toward a network with less capacity).

There are several secondary effects which can occur during the handover. For instance, a reorganization of packets, an RTT increase, which can in turn trigger TCP congestion control operations and further harm the performance. If the TCP congestion control is not able to adapt quickly to the conditions of the new network, after the handover, it can result in the dropping of multiple packets.

No IP mobility framework (network layer method) can help in solving this problem. It must be managed by the transport layer. The TCP sender is not aware of the change of mobile node (MN), and the handover may be perceived as congestion, leading to the triggering of its congestion control. There are two main criteria which lead to secondary effects on the TCP stream control:

– the way in which the handover is carried out at layer 2/3: make-before-break (MBB) or break-before-make (BBM);

– the evolution of the BDP: an upward handover (which means an increase in BDP) or downward (decrease in BDP), and in particular the evolution of the delay.

6.5.3. *TCP impacts for a break before make handover*

The BBM means that layer 2 is interrupted during the network change. The new connection is only established when the old connection is closed. The effects of BBM are summarized below:

– *risk of disconnection* if the duration of disturbances is too long (applications and/or TCP timeout);

– *the dropping of packets* during the interruption can trigger a congestion mechanism even though it is "only" a drop, thus leading to an *inefficient use of bandwidth*;

– *connection time not used*: an RTO may timeout during the interruption and lead to a longer time to recover the communication even after the connection is re-established. Moreover, the RTO could expire several times after as the new network delay could be too long for the actual RTO calculation. The timeout of the RTO(s) prevents the sender from transmitting data as the ssthresh (slow start threshold) value is reduced and the recovery will take longer since it will begin with a slow start with low ssthresh and cwnd values.

6.5.4. *TCP impacts for a make before break handover*

The MBB means that a new layer 2 is prepared before the old layer is deactivated. Two connections are available for a short period at the time of the handover. The effects of this are summarized in Table 6.2.

	Upward Handover	Downward Handover
Impact of the delay	*Useless retransmissions*: the RTO times out because the previous RTT is not big enough to correspond with the new one. The calculation of the RTO cannot adapt to this unexpected increase. *Parasitic RTOs occur as a secondary effect*	*Useless retransmissions*: Some packets "in flight" may be received out of order, if the new connection delivers packets faster than the previous one. Therefore, the receiver interprets this as packets being dropped and triggers a retransmission. *Convergence toward slower RTOs*
Impact of the BDP	*Unused bandwidth and resources wasted*: TCP may not be able to use the higher bandwidth of the new network.	*Real congestion / dropping of packets*: The TCP stream will be too high for the bandwidth and the new network may not be able to deliver the packets. The congestion in the network which occurs is real, packets may be deleted, and the TCP stream control activated. *Parasitic RTOs*

Table 6.2. *Impact of the MBB handover on TCP*

6.5.5. *The effect on TCP of the vertical handover with simultaneous variation of the bandwidth and delay*

In order to illustrate the impact of a handover between a terrestrial and a satellite network over a TCP connection, we configured the simulator used previously in order to model the passage from a satellite network to a terrestrial network with a variation in the delay and bandwidth. The satellite network is simulated by an OWD of 250 ms and a bandwidth of 2 Mbps.

This experiment shows the behavior of TCP versions during handovers of satellite networks to other terrestrial networks, or vice versa. This also helps to extract the best and most suitable protocol for satellite/terrestrial hybrid networks with apparent changes in bandwidth and delay.

There are communication improvements and optimizations, as stated in the beginning of the chapter. The most commonly used solution is the PEP. This provides improvements to transmissions, while "protecting" the terrestrial segment from the characteristics of the satellite network, as well as stream acceleration and optimization mechanisms. However, this also comes with a number of disadvantages, such as incompatibility with encrypted connection and breaks in the end-to-end semantic of a TCP connection. Moreover, PEPs may be detrimental during mobility toward a terrestrial network. The version used for tests is PEPSal.

6.5.5.1. *Configuration and measuring*

The test is initiated at 512 Kbps/500 ms (500 ms of RTT) for the first 20 s, then a handover occurs at t = 20 to a terrestrial network (2 Mbps/50 ms). At t = 40 s, another handover occurs to return to the initial network. The error rate is 0.

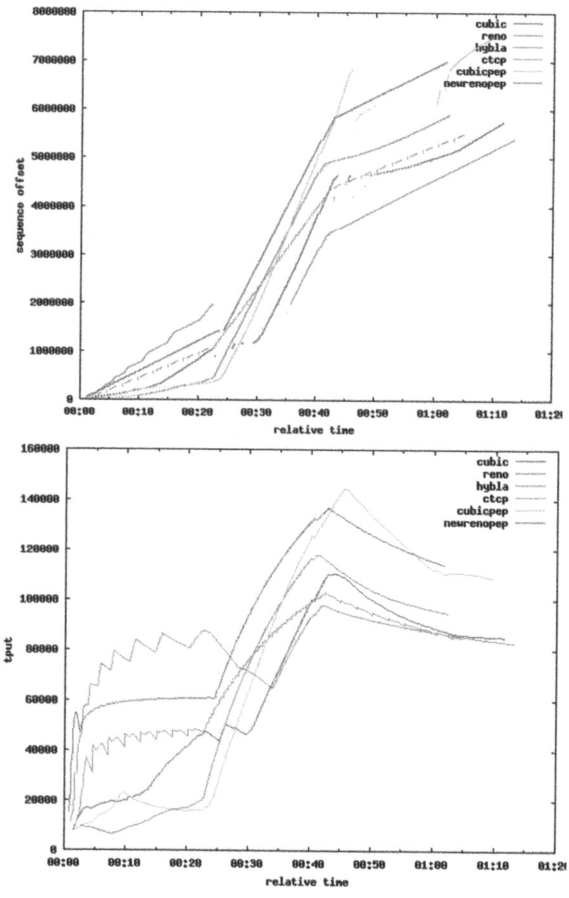

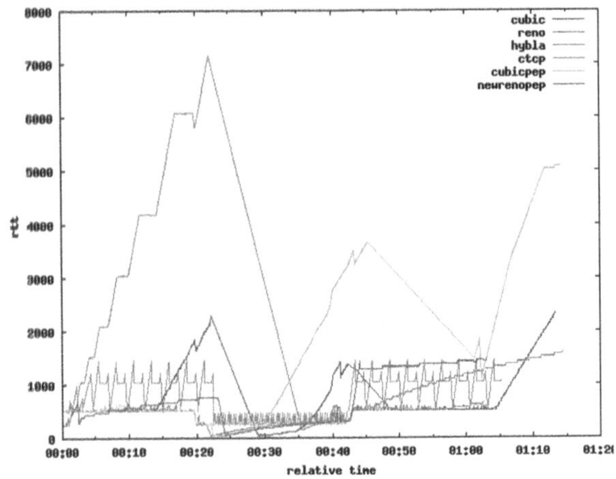

Figure 6.7. *Evolution in sequence numbers, transfer rates and RTTs during a handover. For a color version of the figure, see www.iste.co.uk/berthou/networks.zip*

6.5.5.2. *Analysis*

In Figure 6.7, the evolution of New Reno with PEPsal is very much permissible over the satellite network. A good reaction from Hybla can also be observed in this phase but with a significant break at the first handover. New Reno has poor behavior and confirms its incompatibility with satellite networks. CTCP is stable and offers a transfer rate which is comparable with PEP solutions. The average transfer rate of Hybla is higher but shows small variations. The last graph shows the RTT applied and shows Hybla's buffering problems, thus explaining the break around the 20 s mark.

6.5.6. *Conclusion*

During a handover from a satellite network to a WiFi network, no major problems were encountered provided that the TCP version used is efficient over satellite connections. All TCP versions are able to rapidly use the surplus bandwidth.

In the other direction (a handover from a WiFi network to a satellite network), the fall in the transfer rate does not cause any particular problems. However, the increase in propagation delay does introduce a number of problems, as well as a sudden increase in buffering over the access router.

The reduced transfer rate combined with the increased delay generates drops which are detrimental for protocols which are too aggressive. CTCP and Hybla are not particularly disturbed by the handover, unlike Cubic. The use of PEP can improve the situation when used alongside Hybla (Hybla PEPsal). Nonetheless, we have shown that using CTCP means that a PEP solution is not necessary.

Furthermore, we also showed that CTCP adapts well to changes in delay because it is the only protocol which integrates delay variation in its congestion control. Congestion control based on drops and delays is, therefore, the best solution for solving the problems encountered in the creation of hybrid networks.

However, ECN solves problems of aggressiveness in some versions, such as Cubic and Hybla. Buffering is best managed during the change of a network to a satellite network.

BBM scenarios were also studied. When the break in the connection at the time of the network change lasts 500 ms, CTCP confirms its superiority over the other versions. Indeed, CTCP is more reactive at the restart than Cubic and Hybla. When the break in the connection at the time of the network change lasts 1,000 ms, the difficulty increases for Cubic due to its aggressive nature.

6.6. General conclusion

This chapter has highlighted the general qualities of the Compound version of the TCP protocol (CTCP). CTCP is a good transport level solution for hybrid networks. This chapter has also demonstrated its adaptation to satellite networks and its capacity to support inevitable network changes which occur in satellite/terrestrial hybrid networks. As things currently stand, we believe it is beneficial to use this version for hybrid networks.

Conclusion

The creation of hybrid satellite networks is an issue which is currently at the heart of the concerns of many stakeholders in the field, whether they are operators, research communities, manufacturers, as well as end users who also want a unified, easy-to-access service everywhere all of the time. These hybrid networks already exist to some extent via “triple play” offers from operators which provide Internet access via asymmetric digital subscriber line (ADSL), and television services via satellite when the ADSL transfer rate falls short. However, it is clear that the integration of such systems is not sufficient for offering the new services expected.

Initial studies into 5G, beyond looking for new bands or coding techniques, highlight the need to extend existing services toward greater mobility and transfer rate while targeting new markets such as machine-to-machine (M2M) and communicating objects. To achieve this, the new standard looks to homogeneously integrate existing communication supports, whether they are wireless, cellular, ad hoc, short or long reach, low or high consumption.

Indeed, the satellite has a particularly important role to play in guaranteeing mass broadcasting and access to areas which are difficult to reach. Satellite communication systems provide a wide range of services, such as broadcasting, high transfer rate and narrowband transmission, to fixed, portable or mobile terminals with global or regional coverage. They are known for offering a reliable and ubiquitous service which is financially beneficial for broadcasting and collection. They are, therefore, an ideal complement to other telecommunication technologies.

The integration of satellite telecommunication into 5G is currently being promoted by the "SatCom" working group involving a number of different stakeholders in the sector.

The use of Q/V frequency bands for future high-speed communication systems will also be a key catalyst to the creation of hybrid networks. Indeed, transmission in these bands is highly sensitive to problems linked to the weather. These systems must be equipped with multiple gateways in order to overcome any problems caused by local weather disturbances, which can lead to significant deterioration in a single gateway. With multiple interconnected gateways in a fiber optic loop, the traffic can be immediately rerouted toward another access point in order to ensure a continuity of service.

Nevertheless, management problems arise in this optic/satellite hybrid network, due to the management interfaces in these networks being very different. The first ones are generally operated by multiprotocol label switching (MPLS), the second ones are operated by proprietary applications for the administration of the satellite network. The problem becomes yet more complicated when the operator of the optic loop is different from the satellite system operator where opportunities for interactions are very limited between the two networks. Due to a lack of interfaces, it is difficult to implement dynamic services which enable the best use of the resources available in both networks.

Quality of service (QoS) management in these hybrid networks is, therefore, essential. This book has examined the most cutting-edge solutions for terrestrial and satellite networks. In some cases, more basic and less flexible solutions can be implemented, such as the use of MPLS tunnels in terrestrial networks. However, in any case, there are needs for signaling and management, and internetwork interoperability problems remain open.

Chapter 4, which addresses the integration of satellite networks into an IP multimedia subsystem (IMS) architecture, offers a solution based on open interfaces, such as session initiation protocol (SIP), common open policy service (COPS) and double RADIUS (DIAMETER) protocols. Works carried out in this field have shown the feasibility of such a solution.

Mobility is key in next-generation communication networks but is also a complex service to manage. The designed of hybrid networks once again causes problems related to interfaces between networks, and also in terms of performance, depending on the solution chosen. Two solutions were put forward in the chapter on mobility. The first solution is simpler to implement and is based on the use of the mobile internet protocol: mobile IP. We have shown that this solution is acceptable in terms of performance, while taking care to choose suitable versions for hybrid networks, such as hierarchical versions and rapid handovers. The second solution is more complex and offers the double advantage of providing better performance in terms of interruption time, as well as the possibility of being combined with QoS management mechanisms, such as those presented in the IMS architecture.

Finally, this book addressed the difficult issue of the performance of the transport layer in hybrid systems. At the protocol level, it is no longer a problem of interfaces, but one of performance. Indeed, the transport layer, running "end-to-end" from one extremity of the network to the other, must adapt to any type of support, whether that be terrestrial, satellite, one or the other in the case of mobility, or a mixture of both. Today, complex solutions of proxy servers are implemented to overcome the poor performance of the early versions of Transmission Control Protocol (TCP). However, it appears that it is evolving in a positive way, as we have shown with the analysis of performance of the cubic and compound TCP versions. In compound, the use of a double congestion window, computed with drops and time taken to cross the network, is providing highly satisfactory results in the hybrid network context.

The near future promises new opportunities for the deployment of satellite communication systems and their integration into terrestrial networks.

We can of course cite the evolutions in communication standards, such as the arrival of DVB-S2X, the successor of the standard of the same name. Improvements in efficiency of up to 51% are expected compared to the initial standard, mainly due to an optimization of modulations and signal processing techniques. A range of modulation structures from 8 to 256 APSK has also appeared, with 112 modcods. Low-signal to noise ration (SNR) modcods for mobile applications (terrestrial, maritime and air) are also emerging.

Other changes can be expected in the higher layers, and especially in the web's flagship protocol, HTTP, which is evolving toward the HTTP/2 version. This evolution, brought about by the generalization of terrestrial mobile terminals, may also be beneficial to satellite systems. Indeed, the main evolution is in the ability of the protocol to multiplex the HTTP requests during the consultation of complex documents, thus dramatically reducing loading time compared to the previous version which sequences the requests. The improvement is even more significant given that there is a high network delay. The possibility of anticipating content transfer by the server is another interesting aspect of this change. The server can decide to transfer to the client, without any prior request, the content which it may need to decode a document, for example (typically, cascading style sheets (CSS) sheets).

Today, our attention is focused on the revolution expected to be brought about by software defined networks (SDNs). This new paradigm offers a decoupling of the control and data planes, making it possible to simplify switching equipment in order to move the intelligence into centralized equipment known as controllers. Through the standardization of a new programming interface, this change will enable the deployment of new added-value services by freeing themselves from the proprietary applications of the equipment manufacturers. Therefore, a new routing protocol may be deployed as a simple application on the controller of a network without any modification of its underlying infrastructure. This ability of deploying new protocols and the centralized control of equipment may give rise to new opportunities in the management of hybrid networks. Flexible materials combined with a standardized interface within a hybrid network of the same operator can comfortably enable the configuration of new services from a single controller with the global vision of the network and the traffic. The satellite network, like terrestrial network equipment, must expose this controller to a standard interface, as the openflow protocol currently advised, and behave as simple switching elements. Although there is now a range of equipment compatible with this type of protocol, this is not the case for satellite networks. Their management remains complex and usually proprietary. Efforts must be made to adapt them to this new paradigm. A complementarity between what can be exposed via a unified interface and what is specific to the satellite must be found to improve flexibility without losing the possibilities offered by precise management of the system.

Bibliography

[ALL 00] ALLMAN M. *et al.*, Ongoing TCP research related to satellites, IETF 2760, February 2000.

[ALL 03] ALLMAN M., FALL K., WANG L. *et al.*, A conservative selective acknowledgment (SACK)-based loss recovery algorithm for TCP, IETF RFC 3517, 2003.

[ALL 07] ALLMAN M., JAIN A., SAROLAHTI P. *et al.*, RFC 4782: quick-start for TCP and IP, IETF experimental, 2007.

[ALL 99] ALLMAN M., GLOVER D., SANCHEZ L., Enhancing TCP over satellite channels using standard mechanisms, IETF RFC 2488, January 1999.

[AVR 10] AVRACHENKOV K., AYESTA U., BLANTON J. *et al.*, RFC 5827: early retransmit for TCP and stream control transmission protocol (SCTP), IETF experimental, 2010.

[AWD 99] AWDUCHE D., MALCOLM J., AGOGBUA J. *et al.*, Requirements for traffic engineering over MPLS, RFC 2702 (informational), September 1999.

[BLA 98] BLAKE S. *et al.*, An architecture for differentiated services, RFC 2475, December 1998.

[BOR 01] BORDER J., KOJO M., GRINER J. *et al.*, Performance enhancing proxies intended to mitigate link-related degradations, IETF 3135, June 2001.

[BOY 00] BOYLE J. *et al.*, The COPS (common open policy service) protocol, IETF RFC 2748, January 2000.

[BRA 92] BRADEN R., BORMAN D., JACOBSON V., TCP extensions for high performance, IETF RFC 1323, 1992.

[BRA 94a] BRADEN R., T/TCP – TCP extensions for transactions functional specification, RFC IETF 1644, July 1994.

[BRA 94b] BRADEN R. *et al.*, Integrated services in the internet architecture: an overview, RFC 1633 (IETF), 1994.

[BRA 97] BRADEN R. *et al.*, Resource reservation protocol (RSVP), version 1 functional specification, RFC 2205, September 1997.

[CAL 03] CALHOUN P. *et al.*, RFC 3588: diameter base protocol, IETF, September 2003.

[CAM 02] CAMARILLO G., MARSHALL W., ROSENBERG J., RFC 3312: integration of resource management and session initiation protocol (SIP), IETF RFC 3312, October 2002.

[CAM 03] CAMARILLO G., MONRAD A., Mapping of media streams to resource reservation flows, RFC 3524, April 2003.

[CHA 03] CHAN K., SAHITA R., HAHN S. *et al.*, Differentiated services quality of service policy information base, IETF RFC 3317, March 2003.

[CLA 90] CLARK D.D., TENNENHOUSE D.L., "Architectural considerations for a new generation of protocols", *Proceedings of IEEE Sigcomm (Symposium on Communications Architectures and Protocols*), pp. 200–208, Philadelphia, PA, September 1990.

[COM 10] COMBES S., SatLabs system recommendations – quality of service specifications, ESA/ESTEC, June 2010.

[DAV 02] DAVIE B. *et al.*, An expedited forwarding PHB (per-hop behavior), RFC 2598, March 2002.

[DEV 07] DEVARAPALLI V., DUPONT F., RFC 4877: mobile IPv6 operation with IKEv2 and the revised IPsec architecture, IETF RFC 4877, April 2007.

[DUB 20] DUBOIS E., FASSON J., DONNY C. *et al.*, "Enhancing TCP based communications in mobile satellite scenarios: TCP PEPs issues and solutions", *ASMS 2010*, pp. 476–483, 2010.

[DUR 01] DUROS E., IZUMIYAMA H., FUJII N. *et al.*, RFC 3077: a link-layer tunneling mechanism for unidirectional links, IETF, March 2001.

[EGG 09] EGGERT L., GONT F., TCP user timeout option, IETF RFC 5482, 2009.

[EMS 04] EMS TECHNOLOGIES, VoIP over Satellite, EMS Technologies Canada Technical Notes, Revision 1-2, May 2004.

[FAI 05] FAIRHURST G., COLLINI-NOCKER B., RFC4626: unidirectional lightweight encapsulation (ULE) for transmission of IP datagrams over an MPEG-2 transport stream (TS), IETF, December 2005.

[FAI 07] FAIRHURST G., MONTPETIT M.-J., RFC 4947: address resolution for IP datagrams over MPEG-2 networks, IETF, July 2007.

[FIR 04] FIRRINCIELI R., CAINI C., "TCP Hybla: a TCP enhancement for heterogeneous networks", *International Journal of Satellite Communications and Networking*, vol. 22, 2004.

[FOX 89] FOX R., TCP big window and Nak options, IETF RFC 1106, June 1989.

[GOD 00] GODERIS D. *et al.*, Service level specification semantics, parameters and negotiation requirements, available at: www.ist-tequila.org/standards/draft-tequila-diffserv-sls-00.txt, July 2000.

[GRO 99] GROSSMAN D., HEINANEN J., RFC2684: multiprotocol encapsulation over ATM adaptation layer 5, IETF, September 1999.

[GUN 08] GUNDAVELLI S., LUNG K., DEVARAPALLI V. *et al.*, RFC 5213: proxy mobile IPv6, IETF RFC 5213, August 2008.

[HAN 05] HANCOCK R. *et al.*, Next steps in signaling (NSIS): framework, RFC 4080, June 2005.

[HAN 98] HANDLEY M., JACOBSON V., RFC 2327: SDP: session description protocol, April 1998.

[HEI 99] HEINANEN J. *et al.*, Assured Forwarding PHB Group, RFC 2597, June 1999.

[HER 00] HERZOG S., BOYLE J., COHEN R. *et al.*, COPS usage for RSVP, IETF RFC: 2749, January 2000.

[IPT 05] IPTEL-VIA-SAT, Cookbook for IP Telephony via DVB-RCS, version 1.0, ESA Project report, 12 May 2005.

[ISO 94] ISO/IEC 13818-16, Generic coding of moving pictures and associated audio information, 1994.

[ISO 00a] ISO/IEC 13818-1, Information Technology – Generic Coding of Moving Pictures and Associated Audio Information: Systems, 2nd ed., December 2000.

[ISO 00b] ISO/IEC 13818-6, Information Technology – Generic Coding of Moving Pictures and Associated Audio Information: Part 6: Extensions for DSM-CC, 2nd ed., December 2000.

[ITU 04] An architectural framework for support of quality of service in packet networks, ITU-T Recommendation Y.1291, 2004.

[JAC 90] JACOBSON V., Compressing TCP/IP header for low-speed serial links, IETF RFC 1144, February 1990.

[JAC 92] JACOBSON V., BRADEN R., BORMAN D., TCP extensions for high performance, IETF RFC 1323, May 1992.

[JOH 04] JOHNSON D., PERKINS C., ARKKO J., RFC 3775: mobility support in IPv6, IETF RFC 3775, June 2004.

[KEM 07] KEMPF D., RFC 4830: problem statement for network-based localized mobility management (NETLMM), IETF RFC 4830, April 2007.

[KOH 06] KOHLER E. *et al.*, Datagram congestion control protocol (DCCP), IETF RFC 4340, March 2006.

[KOO 09] KOODLI R., RFC 5568: mobile IPv6 fast handovers, IETF RFC 5568, July 2009.

[KRI 08] KRISTIANSEN E., ROSETI C., "TCP Noordwijk: optimize TCP-based transport over DAMA in satellite networks", *International Communications Satellite Systems Conference*, 2008.

[LEE 09] LEE D., CARPENTER B.E., BROWNLEE N., Observations of UDP to TCP ratio and port numbers, Technical report, Department of Computer Science, University of Auckland, 3 December 2009.

[LEF 02] LE FAUCHEUR F., WU L., DAVIE B. *et al.*, Multi-protocol label switching (MPLS) support of differentiated services, RFC 3270 (proposed standard), May 2002.

[MAH 96] MAHDAVI J., FLOYD S., ROMANOW A. *et al.*, TCP selective acknowledgment options, IETF RFC 2018, 1996.

[MAN 04] MANNER J., KOJO M., RFC 3753: mobility related terminology, IETF RFC 3753, June 2004.

[MAN 05] MANNER J., KARAGIANNIS G., MCDONALD A. *et al.*, NSLP for quality-of-service signaling, Technical report, IETF, 2005.

[MAT 96] MATHIS M., MAHDAVI J., FLOYD S. *et al.*, TCP selective acknowledgment options, IETF RFC 2018, October 1996.

[MOO 01] MOORE B. *et al.*, Policy core information model – specification, RFC 3060, February 2001.

[NER 04] NERA, VoIP over DVB-RCS – a radio resource and QoS perspective, NERA VoIP White paper, December 2004.

[NIC 98] NICHOLS K., BLAKE S., BAKER F. *et al.*, RFC2474 definition of the differentiated services field (DS field) in the IPv4 and IPv6 headers, December 1998.

[NIC 99] NICHOLS K., JACOBSON V., ZHANG L., A two-bit differentiated services architecture for the internet, RFC 2638, July 1999.

[OUR 10] OURSES PROJECT, Aerospace valley, available at http://www.ourses-project.fr/, 2010.

[PAD 02] PADMANABHAN V., FAIRHURST G., SOORIYABANDARA M. *et al.*, RFC 3449, TCP performance implications of network path asymmetry, BCP 69, IETF, December 2002.

[PER 02] PERKINS C., RFC 3344: IP mobility support for IPv4, IETF RFC 3344, August 2002.

[POS 80] POSTEL J., RFC-768: user datagram protocol, request for comments, August 1980.

[POS 81] POSTEL J., RFC 793: transmission control protocol, Technical report, Internet Engineering Task Force (IETF), 1981.

[RAM 01] RAMAKRISHNAN K., FLOYD S., BLACK D., The addition of explicit congestion notification (ECN) to IP, IETF RFC 3168, September 2001.

[RAM 07] RAMOS A., DE LA CUESTA B., CARRO B. *et al.*, "A novel QoS architecture for next generation broadband satellite systems", *International Workshop on IP Networking over Next-Generation Satellite Systems (INNSS '07)*, Budapest, Hungary, 5 July 2007.

[RHE 08] RHEE I., XU L., HA S., "CUBIC: a new TCP-friendly high-speed TCP variant", *SIGOPS Operating Systems Review*, vol. 42, no. 5, pp. 64–74, New York, USA, available at http://doi.acm.org/10.1145/1400097.1400105, July 2008.

[RIG 00a] RIGNEY C. *et al.*, RFC 2865: remote authentication dial in user service (RADIUS), IETF, June 2000.

[RIG 00b] RIGNEY C. *et al.*, RFC 2866: RADIUS accounting, IETF, June 2000.

[ROS 02a] ROSENBERG J., SCHULZRINNE H., CAMARILLO G. *et al.*, SIP: session initiation protocol, IETF RFC-3261, June 2002.

[ROS 02b] ROSENBERG J. *et al.*, An offer/answer model with the session description protocol (SDP), RFC 3264, June 2002.

[SAL 00] SALSANO S. *et al.*, Definition and usage of SLSs in the AQUILA consortium, available at: www.ist-tequila.org/standards/draft-salsano-aquila-sls-00.txt, Internet Draft, November 2000.

[SAL 02] SALSANI S. *et al.*, "QoS control by means of COPS to support SIP-based applications", *IEEE Network*, vol. 16, no. 2, pp. 27–33, March/April 2002.

[SAT 10] SATLABS GROUP, DVB-RCS management information base, February 2010.

[SCH 03] SCHULZRINNE H., CASNER S., FREDERICK R. *et al.*, RTP: a transport protocol for real-time applications, IETF RFC 3550, July 2003.

[SCH 06] SCHULZRINNE H., HANCOCK R., General internet messaging protocol for signaling, Technical report, IETF, 2006.

[SCP 06] SPCS, CCSDS 714.0-B-2 transport protocol, no. 2, October 2006.

[SKI 05] SKINNEMOEN H., VERMESAN A., IUORAS A. *et al.*, VoIP over DVB-RCS with QoS and bandwidth on demand, *Wireless Communications IEEE*, vol. 12, no. 5, pp. 46–53, October 2005.

[SOL 08] SOLIMAN H., CASTELLUCCIA C., EL MALKI K. *et al.*, RFC 5380: hierarchical mobile IPv6 (HMIPv6) mobility management, IETF RFC 5380, October 2008.

[STE 00] STEWART R., RFC 2960: stream control transmission protocol, October 2000.

[STI 05] STIEMERLING M., Loose end message routing method for NATFW NSLP, Technical report, IETF, 2005.

[STM 07] STM SatLink VSAT user guide, STM Norway AS Publication no. 101557, Rev. U, 9 August 2007.

[THO 98] THOMSON S., NARTEN T., JINMEI T., RFC 4862: IPv6 stateless address autoconfiguration, IETF RFC 4862, December 1998.

[WEN 10] WEN J., ZHANG J., HAN Y. *et al.*, "TCP-FIT: an improved TCP congestion control algorithm and its performance", *INFOCOM 2011*, October 2010.

[WES 01] WESTERINEN A. *et al.*, Terminology for policy-based management, RFC 3198, November 2001.

[YAM 08] YAMAMOTO T., "Estimation of the advanced TCP/IP algorithms for long distance collaboration", *Proceedings of the 6th IAEA Technical Meeting on Control, Data Acquisition, and Remote Participation for Fusion Research*, vol. 83, nos. 2–3, pp. 516–519, April 2008.

[YAV 00] YAVATKAR R., PENDARAKIS D., GUERIN R., A framework for policy-based admission control, IETF RFC 2753, January 2000.

[ZHA 06a] ZHANG Q., SRIDHARAN M., KUN T. *et al.*, "A compound TCP approach for high-speed and long distance networks", *INFOCOM*, 2006.

[ZHA 06b] ZHANG Q., SRIDHARAN M., KUN T. *et al.*, "Compound TCP: a scalable and TCP-friendly congestion control for high-speed networks – protocols for fast long-distance networks", *PFLDNet*, 2006.

Standards

[AME 04] EUROPEAN SPACE AGENCY, Interactive Broadband DVB-RCS/S OBP Communication System (AMERHIS), available at http://www.esa.int, 2004.

[ETS 99] ETSI TR 100 815 v1.1.1, Digital video broadcasting (DVB); guidelines for the handling of asynchronous transfer mode (ATM) signals in DVB systems, ETSI Technical report, February 1999.

[ETS 01] ETSI TR 101 984 V1.1.1, Satellite earth stations and systems (SES); broadband satellite multimedia; services and architectures, November 2001.

[ETS 03] ETSI TR 101 202 v1.2.1, Digital video broadcast (DVB); implementation guidelines for data broadcasting, ETSI Technical report, January 2003.

[ETS 04] ETSI EN 301 192 V1.4.1, Digital video broadcast (DVB); DVB specification for data broadcasting, Norme ETSI, November 2004.

[ETS 05a] 3GPP TS 23.228 3rd Generation Partnership Project; technical specification group services and system aspects; IP multimedia subsystem (IMS); stage 2 (Release 7), V7.0.0 (2005-06), 2005.

[ETS 05b] ETSI TR 102 187 V1.1.1, Satellite earth stations and systems (SES); broadband satellite multimedia; overview of BSM families, May 2005.

[ETS 06] ETSI, Satellite earth stations and systems (SES), broadband satellite multimedia, QoS functional architecture, ETSI Technical report, TR 101 462, V1.1.1, December 2006.

[ETS 07a] ETSI TR 101 984, Satellite earth stations and systems (SES); broadband satellite multimedia; services and architectures, V1.2.1, December 2007.

[ETS 07b] ETSI TS 29 208, End-to-end quality of service (QoS) signalling flows, ETSI 3GPP, June 2007.

[ETS 07c] ETSI TS 102 606, Digital video broadcasting (DVB); generic stream encapsulation (GSE) protocol, 2007.

[ETS 08] ETSI TS 123 228, IP multimedia subsystem (IMS); stage 2, ETSI 3GPP, April 2008.

[ETS 09a] ETSI EN 301 790 v1.5.1, Digital video broadcast (DVB); interaction channel for satellite distribution systems, Norme ETSI, May 2009.

[ETS 09b] ETSI EN 302 307 v1.2.1, Digital video broadcasting, second generation framing structure, channel coding and modulation systems for broadcasting, interactive services, news gathering and other broadband satellite application, Norme ETSI, August 2009.

[ETS 09c] ETSI TR 101 790 v1.4.1, Digital video broadcast (DVB); interaction channel for satellite distribution systems; guidelines for the use of EN 301 790, ETSI, July 2009.

[ETS 09d] ETSI TR 102 676, Satellite earth stations and systems (SES); broadband satellite multimedia (BSM); performance enhancing proxies (PEPs), V1.1.1, November 2009.

[ETS 09e] ETSI TS 102 602, Connection control protocol (C2P) for DVB-RCS; specifications, version 1.1.1, ETSI BSM, January 2009.

[ETS 09f] ETSI TS 102 603, Connection control protocol (C2P) for DVB-RCS; background information, version 1.1.1, ETSI BSM, January 2009.

[ETS 09g] ETSI EN 302 307, Digital video broadcasting (DVB); second generation framing structure, channel coding and modulation systems for broadcasting, interactive services, news gathering and other broadband satellite applications (DVB-S2), 2009.

[ETS 11a] ETSI TS 23 207, End-to-end quality of service (QoS) concept and architecture, ETSI 3GPP, version 10.0.0, March 2011.

[ETS 11b] ETSI TS 102 771, Digital video broadcasting (DVB); generic stream encapsulation (GSE) implementation guidelines, 2011.

[ETS 12a] ETSI EN 300 468, Digital video broadcasting; specification for service information in DVB systems, May 2012.

[ETS 12b] ETSI TS 23 107, Quality of service (QoS) concept and architecture, ETSI 3GPP, version 11, June 2012.

[ETS 12c] ETSI EN 301 545 (1 to 3), 2012-5, Digital video broadcasting (DVB); second generation DVB interactive satellite system (DVB-RCS2); 1 – overview and system level specification; 2 – lower layers for satellite standard; 3 – higher layers satellite specification, 2012.

Index

Other titles from

iSTE

in

Networks and Telecommunications

2015

BENSLAMA Malek, KIAMOUCHE Wassila, BATATIA Hadj
Connections Management Strategies in Satellite Cellular Networks

BENSLAMA Malek, BATATIA Hadj, BOUCENNA Mohamed Lamine
Ad Hoc Networks Telecommunications and Game Theory

PUJOLLE Guy
Software Networks

2014

ANJUM Bushra, PERROS Harry
Bandwidth Allocation for Video under Quality of Service Constraints

BATTU Daniel
New Telecom Networks: Enterprises and Security

BEN MAHMOUD Mohamed Slim, GUERBER Christophe, LARRIEU Nicolas, PIROVANO Alain, RADZIK José
Aeronautical Air−Ground Data Link Communications

BITAM Salim, MELLOUK Abdelhamid
Bio-inspired Routing Protocols for Vehicular Ad-Hoc Networks

CAMPISTA Miguel Elias Mitre, RUBINSTEIN Marcelo Gonçalves
Advanced Routing Protocols for Wireless Networks

CHETTO Maryline
Real-time Systems Scheduling 1: Fundamentals
Real-time Systems Scheduling 2: Focuses

EXPOSITO Ernesto, DIOP Codé
Smart SOA Platforms in Cloud Computing Architectures

MELLOUK Abdelhamid, CUADRA-SANCHEZ Antonio
Quality of Experience Engineering for Customer Added Value Services

OTEAFY Sharief M.A., HASSANEIN Hossam S.
Dynamic Wireless Sensor Networks

PEREZ André
Network Security

PERRET Etienne
Radio Frequency Identification and Sensors: From RFID to Chipless RFID

REMY Jean-Gabriel, LETAMENDIA Charlotte
LTE Standards
LTE Services

TANWIR Savera, PERROS Harry
VBR Video Traffic Models

VAN METER Rodney
Quantum Networking

XIONG Kaiqi
Resource Optimization and Security for Cloud Services

2013

ASSING Dominique, CALÉ Stéphane
Mobile Access Safety: Beyond BYOD

BEN MAHMOUD Mohamed Slim, LARRIEU Nicolas, PIROVANO Alain
Risk Propagation Assessment for Network Security: Application to Airport Communication Network Design

BEYLOT André-Luc, LABIOD Houda
Vehicular Networks: Models and Algorithms

BRITO Gabriel M., VELLOSO Pedro Braconnot, MORAES Igor M.
Information-Centric Networks: A New Paradigm for the Internet

BERTIN Emmanuel, CRESPI Noël
Architecture and Governance for Communication Services

DEUFF Dominique, COSQUER Mathilde
User-Centered Agile Method

DUARTE Otto Carlos, PUJOLLE Guy
Virtual Networks: Pluralistic Approach for the Next Generation of Internet

FOWLER Scott A., MELLOUK Abdelhamid, YAMADA Naomi
LTE-Advanced DRX Mechanism for Power Saving

JOBERT Sébastien *et al.*
Synchronous Ethernet and IEEE 1588 in Telecoms: Next Generation Synchronization Networks

MELLOUK Abdelhamid, HOCEINI Said, TRAN Hai Anh
Quality-of-Experience for Multimedia: Application to Content Delivery Network Architecture

NAIT-SIDI-MOH Ahmed, BAKHOUYA Mohamed, GABER Jaafar, WACK Maxime
Geopositioning and Mobility

PEREZ André
Voice over LTE: EPS and IMS Networks

2012

AL AGHA Khaldoun
Network Coding

BOUCHET Olivier
Wireless Optical Communications

DECREUSEFOND Laurent, MOYAL Pascal
Stochastic Modeling and Analysis of Telecoms Networks

DUFOUR Jean-Yves
Intelligent Video Surveillance Systems

EXPOSITO Ernesto
Advanced Transport Protocols: Designing the Next Generation

JUMIRA Oswald, ZEADALLY Sherali
Energy Efficiency in Wireless Networks

KRIEF Francine
Green Networking

PEREZ André
Mobile Networks Architecture

2011

BONALD Thomas, FEUILLET Mathieu
Network Performance Analysis

CARBOU Romain, DIAZ Michel, EXPOSITO Ernesto, ROMAN Rodrigo
Digital Home Networking

CHABANNE Hervé, URIEN Pascal, SUSINI Jean-Ferdinand
RFID and the Internet of Things

GARDUNO David, DIAZ Michel
Communicating Systems with UML 2: Modeling and Analysis of Network Protocols

LAHEURTE Jean-Marc
Compact Antennas for Wireless Communications and Terminals: Theory and Design

RÉMY Jean-Gabriel, LETAMENDIA Charlotte
Home Area Networks and IPTV

PALICOT Jacques
Radio Engineering: From Software Radio to Cognitive Radio

PEREZ André
IP, Ethernet and MPLS Networks: Resource and Fault Management

TOUTAIN Laurent, MINABURO Ana
Local Networks and the Internet: From Protocols to Interconnection

2010

CHAOUCHI Hakima
The Internet of Things

FRIKHA Mounir
Ad Hoc Networks: Routing, QoS and Optimization

KRIEF Francine
Communicating Embedded Systems / Network Applications

2009

CHAOUCHI Hakima, MAKNAVICIUS Maryline
Wireless and Mobile Network Security

VIVIER Emmanuelle
Radio Resources Management in WiMAX

2008

CHADUC Jean-Marc, POGOREL Gérard
The Radio Spectrum

GAÏTI Dominique
Autonomic Networks

LABIOD Houda
Wireless Ad Hoc and Sensor Networks

LECOY Pierre
Fiber-optic Communications

MELLOUK Abdelhamid
End-to-End Quality of Service Engineering in Next Generation Heterogeneous Networks

PAGANI Pascal *et al.*
Ultra-wideband Radio Propagation Channel

2007

BENSLIMANE Abderrahim
Multimedia Multicast on the Internet

PUJOLLE Guy
Management, Control and Evolution of IP Networks

SANCHEZ Javier, THIOUNE Mamadou
UMTS

VIVIER Guillaume
Reconfigurable Mobile Radio Systems

Printed and bound by CPI Group (UK) Ltd, Croydon, CR0 4YY
13/09/2026
14961081-0003